Modeling and Simulation of Mineral Processing Systems

Modeling and Simulation of Mineral Processing Systems

Editor

Mineekshi Awasthi

scitus
academics

Modeling and Simulation of Mineral Processing Systems

Edited by **Mineekshi Awasthi**

Printed in 2017

ISBN: 978-1-68117-491-4

Library of Congress Control Number: 2015936598

© 2016 by
SCITUS Academics LLC,
616, Corporate Way, Suite 2, 4766,
Valley Cottage, NY 10989

www.scitusacademics.com

Contents

Preface

The field of quantitative modeling of mineral processing equipment and the use of these models to simulate the actual behavior of ore dressing and coal washing as they are configured to work in industrial practice. The material is presented in a pedagogical style that is particularly suitable for readers who wish to learn the wide variety of modeling methods that have evolved in this field. The models vary widely from one unit type to another. Wherever possible model structure is related to the underlying physical processes that govern the behaviour of particulate material in the processing equipment. Predictive models are emphasised throughout so that, when combined, they can be used to simulate the operation of complex mineral processing flow sheets. The development of successful simulation techniques is a major objective of the work that is covered in the text.

Editor

UML Profile for Mining Process: Supporting Modeling and Simulation Based on Metamodels of Activity Diagram

Andrea Giubergia[1], Daniel Riesco[2], Verónica Gil-Costa[1], and Marcela Printista[2]

[1]Mine Department, Chacabuco 569, National University of San Luis, 5700 San Luis, Argentina

[2]Informatic Department, Ejército de los Andes 950, National University of San Luis, 5700 San Luis, Argentina

ABSTRACT

An UML profile describes lightweight extension mechanism to the UML by defining custom stereotypes, tagged values, and constraints. They are used to adapt UML metamodel to different platforms and domains. In this paper we present an UML profile for models supporting event driving simulation. In particular, we use the Arena simulation tool and we focus on the mining process domain. Profiles provide an easy way to obtain well-defined specifications, regulated by the Object Management Group (OMG). They can be used as a presimulation technique to obtain solid models for the mining industry. In this work we present a new profile to extend the UML metamodel; in particular we focus on the activity diagram. This extended model is applied to an industry problem involving loading and transportation of minerals in the field of mining process.

INTRODUCTION

The literature on formal use of Unified Modeling Language (UML) [1] is huge and growing rapidly. This tool becomes a modeling standard in 1990 to follow what changed the world of software [2]. UML makes it possible to model any system from different perspectives. Modeling a system from different perspectives allows to clearly picking the vision of what you want to do. This is because UML is a language that holds its own specific rules, semantic, and syntax.

UML can be used as documentation of code, but it is also intended as a means of specifying a system. Model Driven Architecture (MDA) [3] calls for complete systems to be generated automatically from UML models. If the language is not precisely defined, the generated system may not be what the model creators intended.

UML has an abstract syntax which is processed by model transformation and code generation. The relation in UML between concrete diagrammatic syntax and the abstract syntax it represents is complicated enough to be a potential source of error. Precisely defining this relationship could simplify the creation of graphical model editors and facilitate animations [4] and reverse engineering. The definition should clearly delineate concrete syntax, abstract syntax, and semantics, and it should also specify the relationships between these parts.

UML defines several types of diagrams to view the dynamic aspects of a system. One of these diagram types is the activity diagram [5] which is used in this work to document the workflow in a system. UML Diagrams are suitable for system analysis, design, and development. However, UML and its diagrams cannot accurately specify concepts for particular domains. That is why UML embodies the concept of profiles. Profiles are defined by the Object Management Group (OMG) in its specification [1], as a predefined set of stereotypes, tagged values, constraints (by using Object Constraint Language (OCL)), and notation icons that collectively specialize and tailor UML to a specific domain or process. When these elements are introduced the model can be clearly visualized and software developers can improve communication and establish a common vocabulary. Also profiles allow adding information to the model to transform it to other models. Object Constraint Language (OCL) constraints are semantic restrictions added to UML elements. The advantage of profiles is that most UML tools can easily apply them. When using profiles it is not necessary to define neither a special notation nor special tools (UML tool is used).

The UML metamodel (which consists of entities and relationships) of the domain is a process-oriented tool. It is graphically structured for the construction of diagrams or flowcharts. These diagrams show the number of steps required by an entity as it moves into the system.

Another powerful tool for system analysis and design is simulation. Its benefits are applicable to almost all kinds of industry. It involves designing a model of a particular system to solve it by means of a numerical technique (algorithm) and the subsequent execution of a series of experiments with the aim of understanding the behavior of such system under certain conditions [6]. The model should be able to reproduce the actual process behavior as accurately as possible. There are many simulation softwares of general purpose [7–10]; in this work we focus on the Arena [10] simulation because it has become the market-leading discrete event simulation software.

To reconcile the two ways of approaching the same subject, namely, UML and simulation, in this work we propose to model a system of loading and transportation of material in the field of the mining industry. In particular, we propose to extend the UML metamodel using profiles. The objective is to adapt the UML metamodel to be able to represent the basic modules of Arena.

The advantage of the proposed profile is that (a) it allows reusing the stereotypes created for the Arena simulation software. Moreover, those stereotypes can be applied to any other process-oriented simulation software. We can also use them or integrate them as part of any other process-oriented simulation software.

This paper is organized as follows. Section 2 presents related work. The Arena simulation software is briefly described in Section 3. Section

4 presents our mining process case of study. Section 5 presents the proposed profile implementing a simulation environment and its OCL constraints. Conclusions are included in Section 6.

RELATED WORK

Modeling and simulation techniques provide the possibility of studying new strategies and to predict the effect of new policies, new designs, and new strategies which would otherwise be too expensive or even impossible to implement and test on real cases. Both the UML modeling and the simulation approach model from two different communities (or point of view) but they can be used for similar purposes.

Some recent methodologies and techniques for modeling and simulation have been presented in [11–13]. In particular, the authors in [11] use genetic algorithms to improve component analysis. This technique is applied to simulated data collected from the Tennessee Eastman chemical plant. The work in [12] evaluates basic data-driven methods for process monitoring and fault diagnosis. Also in this case the benchmark of Tennessee Eastman (TE) process is utilized to illustrate the efficiencies of the discussed methods. Finally, in [13] the authors present a fault tolerant control architecture which can be reconfigured online. The proposed scheme is evaluated with the TE benchmark model.

In this work we focus on UML modeling and simulation techniques which are widely used in systems engineering. Although these two methodologies have evolved separately, there are some works [14–17] that integrate modeling tools with specific simulation software.

The works presented in [14, 15] show how the UML-Arena combination is used to create a model. Then this model is reused into the simulation environment of the Arena software. In [14] the authors describe the use of activity diagrams to automatically generate simulation models and propose the use of an algorithm as a solution to automatically transform UML models (only activity diagrams) into simulation models. Similarly, the authors in [15] show the use of UML activity diagrams as the basis for building simulation models, which are then run on the Arena simulation software.

In other words, these studies show a methodology to transform activity diagrams into simulation models for Arena. However, they do not take advantage of the UML profiles. A profile created for a specific situation can be reused into other environments with similar characteristics, thus saving time and increasing the versatility of the profile created.

As explained in [18], OMG defines two possible approaches for defining domain specific languages. In the first one, a new alternative language is defined [19]. A new tailor-made language will produce suitable specific notation that will match the concepts of a specific application domain. However, as the new language does not respect UML semantics, it will not allow the use of commercial UML tools for drawing diagrams, generating code, and so forth [18]. The second approach (named Profiles) uses the UML metamodel respecting the original semantic of the UML elements (classes, attributes, etc.). But new constrains are added to the definition and to the relationships of those elements.

The works in [20–22] show the versatility of profiles to define different environments (application domains), which extend the UML specification (not only the syntax but also the semantic through formal OCL constraints). In particular, the work in [20] shows the way for defining design patterns with profile, proposing architecture in levels. It shows how the definition of a profile for a particular pattern, and how an UML tool can be enough to introduce profile for patterns. It analyzes the advantages of using profiles to define, document, and visualize design patterns.

The work in [22] shows the definition of the C&K metrics applied to Aspect Oriented Design (AOD) using UML profiles. A new definition of metrics using profiles without modifying the OMG metamodels is specified. An OCL formal specification is developed. The defined profile allows applying methodologies which use AOD and allows measuring the design. The calculation of the resources is an activity that can improve the software development. The engineers can import this profile to any UML tool and measure the quality of its AOD. They do not need to build a specific tool for AOD. They can use any UML tools and thereby improve the quality of aspect oriented developments.

In [21], the creation of components of the Java EE 6 business platform from technical business processes modeled with Business

Process Model Notation (BPMN) 2.0 is proposed. This generation was achieved by performing three transformations in the context of Directed Model Architecture (MDA). First, a technical model BPMN 2.0 is transformed into an UML class model. Then the class model is transformed into a model with Enterprise Edition (Java EE) profiles. Finally the last generated model is transformed through Meta Object Facility Script (MOFScript) into Java EE components. The transformations are performed with Query View Transformations (QVT) Relations and MOFScript. This work contributes with transformations between profiles generated with Java EE business components related to business processes, so it helps in improving the development productivity and in reducing design errors.

In [16], UML is proposed as an effective structure for building simulation models of hybrid manufacturing systems, where machines communicate through a network or using buffers. The model is implemented with the MATLAB language [23]. UML is used for modeling, but UML profiles are not considered.

The work in [24] presents research results related to the main steps of the modeling and simulation process (e.g., design of simulation experiments), as well as to the application of simulation for solving different problems, inherent to the operation of complex systems (e.g., analysis, optimization, and management). In particular, the authors describe a tool which integrates simulation methodologies that incorporate simulation with other scientific approaches for the analysis, optimization, and management of complex systems. This tool allows for transforming UML models into a simulation environment; in particular the Arena simulation environment is used.

UML has also been used in collaboration with other simulation softwares like DEVS [25] or Petri Nets. In [26], the authors present UML diagrams that are extended with stochastic attributes. This provides possibilities for further simulation of a developed model using a simulation tool called DEVS [25]. The authors generate UML models using use case and activity diagrams. On the other hand, the work in [17] proposes a methodology for transferring knowledge to students. The authors propose to combine the use of Petri Net and UML to model a business process as a system of discrete events. None of these papers create UML profiles.

In this paper we focus on the use of UML activity diagrams such as a presimulation techniques, because they allow capturing dynamic aspects of the behavior of a system. We extend the UML metamodel by defining profiles for a particular domain. Unlike previous work, like described in [14, 15], our work ensures that the UML metamodel is not modified and it also meets all the semantic. Moreover, our proposed profile is specified with the formality given by the OCL [27] language, allowing (1) to reuse the proposed stereotypes with any other simulation software and (2) to reuse the simulation model with any tool that includes profiles.

SIMULATION WITH ARENA

Modeling and simulation provide the basis for the efficient solving of various problems related to the operation of complex systems like analysis, optimization, and management and industry problems like mining process, and so forth. Simulation is considered to be one of the most effective technologies for the analysis and planning of logistics systems.

Using simulation, you can include randomness through properly identified probability distributions taken directly from study data. For example, in this work we consider the loading and transportation of material in the field of the mining industry. In this case, there may be intersections on roads or narrow paths that only allow a truck to move at a time. Simulation also allows including the analysis and evaluation of the system when failures occur.

Other techniques such as spreadsheet analysis or linear, goal, and dynamic programming are useful to maximize or minimize a single element (e.g., cost, utilization, or wait time). But these techniques limit the analysis to only one element, often at the expense of secondary goals. They do not allow randomness. In particular, spreadsheet analysis forces to use the average time and will not be able to accurately capture the variability that exists in reality.

Arena simulation software [10] is a general propose simulation tool enabling the construction of models over a series of modules or basic components organized hierarchically. The Arena simulation software has high level of modeling supporting graphical design. It also includes a lower level of modeling including specific details as arrival times, service times, scheduling of processes, and so forth.

A model is developed using modules that are part of the basic processes. In Arena, modules are the flowchart and data objects that define the process to be simulated. All information required to simulate a process is stored in modules. The dynamics associated with the processes can be viewed as nodes in a network by which entities circulate causing a change in the system state. The entities with attributes and variables compete for the services provided by the resources. Entities are items (like trucks, mineral, etc.) that are being served or produced.

Figure 1 shows a simple model build with Arena. The CREATE module is the starting point for entities in a simulation model. Entities are created using a schedule or based on a time between arrivals. Entities then leave the module to begin processing through the system. The entity type is specified in this module. The PROCESS modules are intended as the main processing method in the simulation. They include the resource by which entities compete. A resource retained by an entity must be released at some point in the model. Otherwise, a deadlock can occur. The DECIDE module allows for decision-making processes in the system. Finally, the DISPOSE modules are ending point for entities in a simulation model.

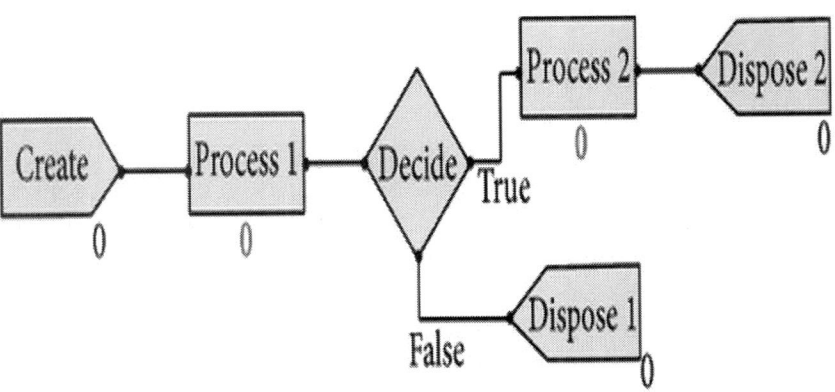

Figure 1: Basic modules used in Arena.

After the model is built, we can run simulations to obtain different metrics and statistics like resources utilization, waiting times, and so forth.

CASE STUDY: MINING PROCESS

A real system model is fundamental when making crucial decisions. A system model offers the possibility of planning with less error. But more importantly, it is possible to investigate new operating procedures that can be used repeatedly without interfering in the operational and daily activities of the system.

The profile proposed in this work is applied to a system of load and transportation of minerals. During the mining process, once the rock or mineral has been removed from its extraction site, it needs to be transported immediately to its final destination. Generally, the transport system consists of a set of trucks.

Several factors are considered for the selection of trucks, the most important factor is the capacity of the vehicle in the mine. To meet daily production requirements is necessary: (a) a good performance of the equipment, that is, to achieve the perfect match "shovel-truck" to obtain maximum extraction and transportation of minerals and (b) the road conditions should be well maintained.

The tasks performed in the daily cycle of a mining process [28, 29] start with the definition of a drilling mesh (according to height, quality, and quantity of the field to be explode). Then, holes are drilled into the field. Then the explosives are introduced into the hole for blasting. The aim is to detach the rock out of the field and crush it. Then blasted material is loaded on the trucks and transported to a discharge area.

Mining Process Simulation with Arena

In this section we explain how to perform the design of load and transportation of minerals in the field of mining process using the Arena simulation tool. However, we emphasize that the focus of this paper is not the simulation design/evaluation of the mining operations but the profile design for discrete event simulations applied to the field of mining projects.

We model a transportation system for a limestone quarry. The trucks move between their loading points to their discharge points using default routes. The cycle begins with the release of trucks («Create») to operating fronts. Trucks are distributed among three different zones (using the module «Decide») to load the mineral. The distribution percentage of the «Decide» module is based on the quality and quantity of material. Once in the loading zone («Process»), if the blade («Resource») is free then the truck is loaded without delay. Otherwise, it should be queued («Queue») until the resource is released.

When the loading process ends, the truck is directed to the discharge zone («Process»), which may be inside the plant or in the sterile area, according to the specific material being transported. This cycle is repeated according to the scheduled work.

Tables 1 and 2 present time involved in a limestone quarry to move trucks from/to the treatment plant to/from the operating fronts. Trucks unload the mineral into a chute in the treatment plant. At the operating fronts we find shovels to load the trucks. Unloaded trucks move faster than loaded trucks. Then the time required to move unloaded trucks from the treatment plant to the operating fronts is smaller than the time required to move the loaded trucks from the operating fronts to the treatment plant as shown in Table 1. In Table 2 we show that the time to load/unload a truck follows an exponential distribution.

Table 1: Time (minutes) for mineral hauling

Route	Shovel 1	Shovel 2	Shovel 3
From plant to solvers	3	5	10
From solvers to plant	8	10	20

Table 2: Time (minutes) for load/unload the mineral to/from the truck

Shovel 1	Shovel 2	Shovel 3	Treatment plant
Exp(6.3)	Exp(6.2)	Exp(6.4)	Exp(5.6)

Figure 2 shows the simulation design implemented with Arena. In this particular case, we evaluate the transportation system of a limestone quarry with tree operation fronts. We set the maximum number of trucks to 5. The number of trucks has been chosen to avoid long queuing times. From each operation front the trucks are loaded with minerals with different "low values," namely, with different quality of the mineral. In each operation front there is a shovel in charge of putting the mineral in the truck. To decide the route for each truck we use an "N-way by chance" «Decide» module and we set the percentage of each route to 33%.

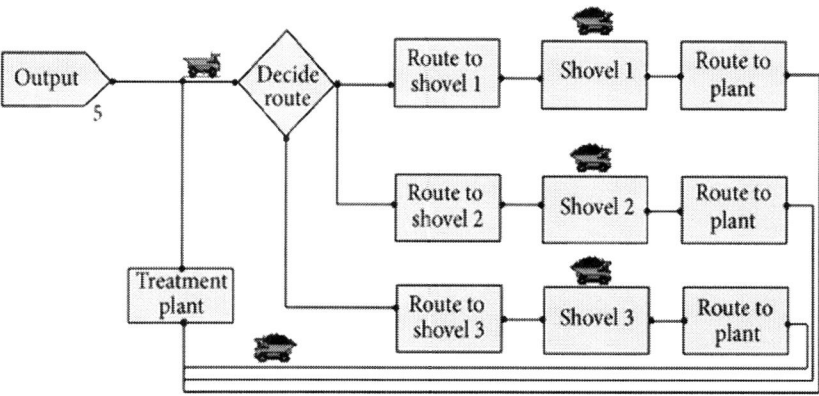

Figure 2: Limestone quarry simulation.

Table 3 shows results obtained after we run the simulation for 20 days (each day has 16 working hours). As we said before the number of trucks is set to avoid long queuing times. Therefore results show that the average number of waiting trucks in each «Resource» (treatment plant, shovels 1, 2, and 3) is less than 1 and the queuing time in the worst case is less than 1 minute.

Table 3: Simulation results

Resource	N° waiting	Avg. waiting time	Max. waiting time
Treatment plant	0.81	0.102	0.889
Shovel 1	0.06	0.024	0.729
Shovel 2	0.08	0.032	0.684
Shovel 3	0.07	0.027	0.596

UML PROFILE DEFINITION FOR MINING PROCESS SIMULATION

UML uses a mechanism called profile [1] to adapt a model to a particular domain. A profile includes three elements: stereotypes, tagged values, and constraints. Stereotypes extend the vocabulary of UML and it is possible to associate tagged values (attributes associated with extended elements) and restrictions [20, 22]. These extension mechanisms allow to adapt a particular domain to an existing metamodel.

The main builder of the profile is the stereotype, indicated as ≪stereotype≫ [1]. It has the same structure or elements (attributes, associations, and operations) of the UML metamodel. So, it is not allowed to modify the semantics, structure, and original concepts [30] of the standard UML.

The proposed stereotypes are related to the components of the activity diagram [5], because they easily adapt themselves to characterize the simulation model [31, 32]. This diagram already has established appropriate notations and concepts showing the steps (activities), decision points, and bifurcations that occur in a transaction so that makes it easier to display. In particular in this work we used the Rational Software Architect (RSA) tool [33] which has the advantage of supporting OCL restrictions. But any tool supporting UML profiles can be used.

The semantic analysis performed to obtain the corresponding equivalences allows identifying the classes of the activity diagram that are necessary to extend through stereotypes. The simulation software has a greater number of modules, besides those mentioned here, some of which are identified with other classes of UML. The elements that are part of the profile are a subset of the UML metamodel applied to a simulation environment. Therefore, the profile presented in this work extends the base classes: InitialNode, Action, DecisionNode, ActivityFinalNode, JoinNode, and ObjectNode ForkNode belonging to the activity diagram.

Although, every stereotype has its own properties, the user can edit the values of those properties. Figure 3shows a slice of the loading-transportation cycle, which highlights the properties of the stereotype ≪Create≫ and its corresponding values.

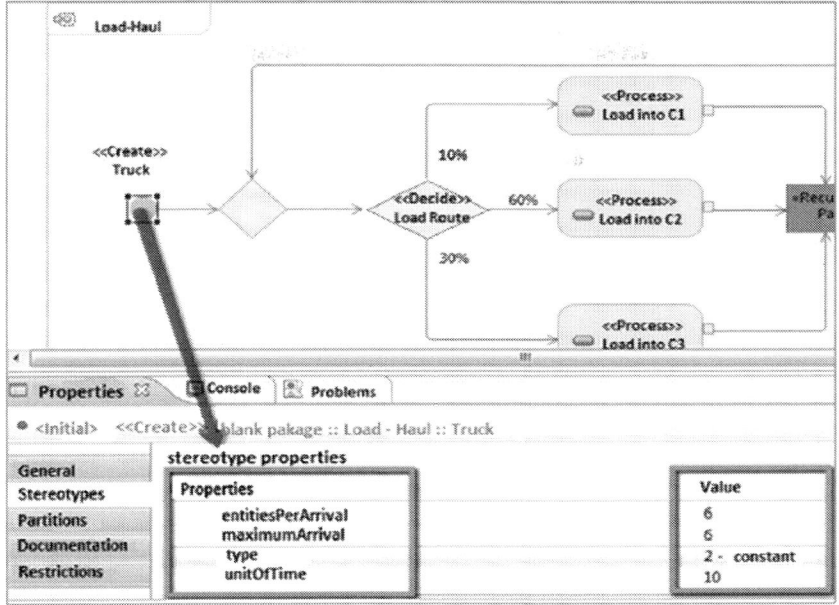

Figure 3: Properties of the stereotype «Create».

The stereotype «Create» extends the InitialNode metaclass. An InitialNode is a starting point for the implementation of an activity. In the UML specification, Superstructure v2.1.2 [5], the InitialNode is defined as a generalization of a control node where an object begins to flow through the system when an activity is invoked. This package belongs to the Basic Activities defined in the abstract syntax of UML Kernel Metamodel of OMG [5]. The Create module has attributes that define the input parameters, which must be defined by the user when starting the simulation. Figure 4 shows the stereotype and the extended metaclass. At the starting point of a simulation model we consider the (a) incoming flow rate generated (random or programmed constant) as well as the (b) number of entities that arrive to the system per unit time. In the mining process described in this work, every truck that enters into the system is a starting point.

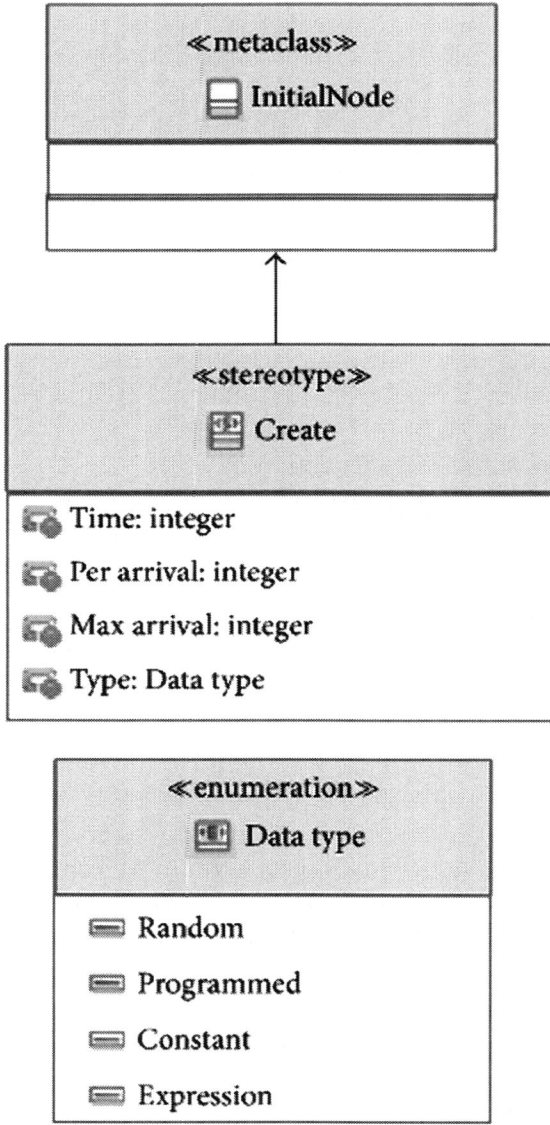

Figure 4: Relationship between metaclass and stereotype «Create».

Figure 5 shows the metamodel corresponding to the OMG specification. In this figure we show the metaclass named InitialNode which is a generalization of the metaclass named ControlNode.

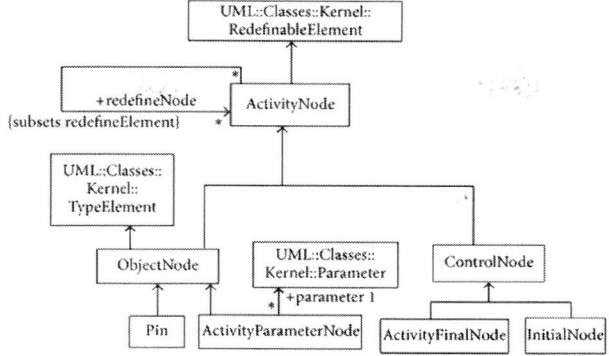

Figure 5: Metamodel of the OMG specification.

The stereotypes ≪Process≫, ≪Assign≫, and ≪Register≫ have a base class named Action [5] in the UML metamodel. For example, if we use an element of type ≪Process≫ in the UML model, it means that the stereotype ≪Process≫ behaves similarly to its base class. Additionally, it has to satisfy the semantics specified with OCL. Figure 6 shows the metaclass named ≪Action≫ and its corresponding stereotypes. For our particular case of study about mining processes related to load and transportation, the ≪Process≫ stereotype represents the load operation by means of the shovel.

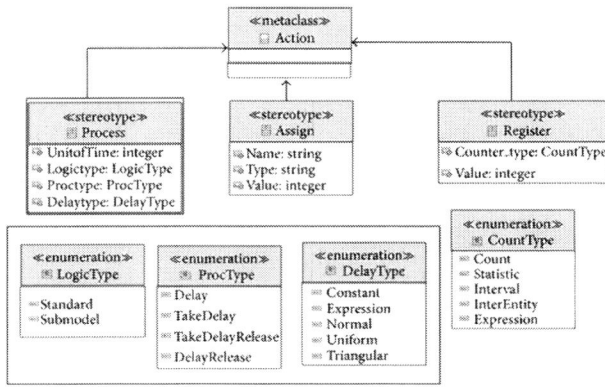

Figure 6: The metaclass action and its stereotypes.

The UML class named JoinNode [5] is extended through a so-called ≪Group≫ stereotype as shown in Figure 7. In a simulation model, this class should specify the number of entities to be clustered as well as how these entities will be grouped (refers to whether to use random groups or, on the contrary, the entities should be grouped according to a specific attribute).

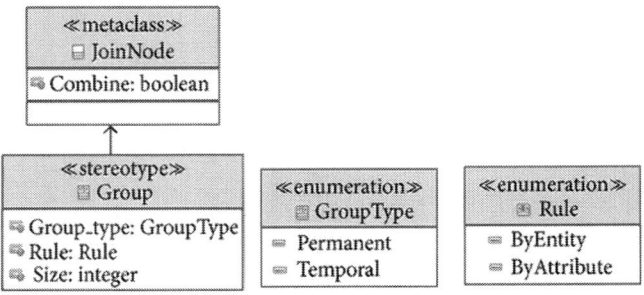

Figure 7: Metaclass JoinNode and its corresponding stereotype.

The class named ObjectNode is extended through four stereotypes ≪Entity≫, ≪WaitingQueue≫, ≪Resource≫, and ≪Set≫ as shown in Figure 8. The base class named ActivityFinalNode of the UML metamodel [5] was extended by a stereotype called ≪Dispose≫. The base class named ForkNode is extended through a stereotype called ≪Divide≫ [5]. This is shown in Figure 9.

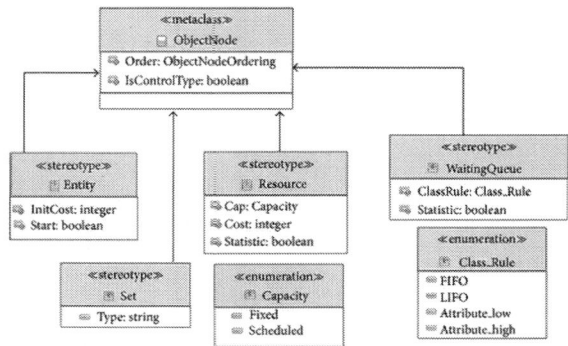

Figure 8: The Object Node metaclass with its stereotypes.

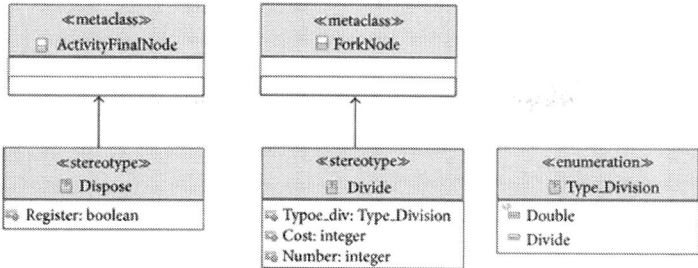

Figure 9: Metaclass ActivityFinalNode and ForkNode with its extended stereotypes.

The stereotype «Decide» [5] extends the UML metaclass named DecisionNode. Figure 10 shows this relationship.

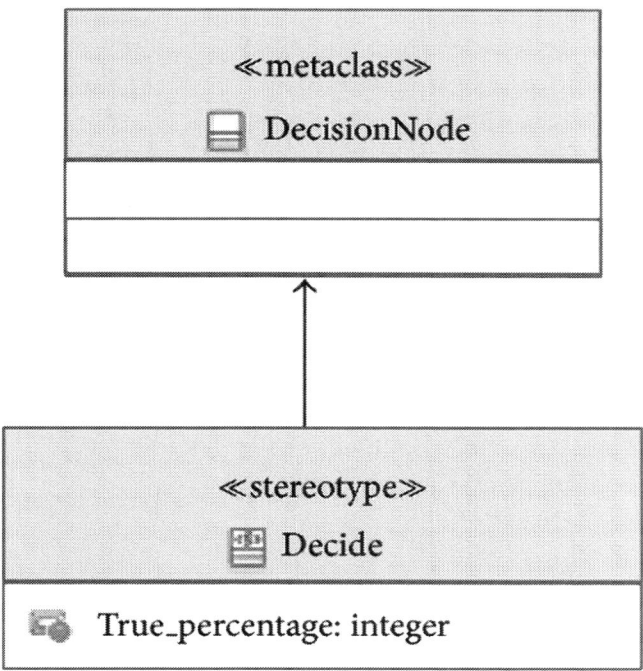

Figure 10: The metaclass Decision Node and its stereotype.

The basic modules that integrate the simulation model of Arena are related to each other through connectors [10]. These connectors indicate how entities flow within the model. These connectors are represented with the metaclass of the activity diagram named ControlFlow. It is not necessary to extend this base class because there is a bilateral matching of the semantics.

Stereotypes defined above integrate the simulation profile that is applied in a mining process. The extended metaclasses were obtained as a result of observing the behavior of the basic modules of the simulation process. This analysis showed that while they share similarities in semantics, these modules have specific attributes that are not specified in the UML base classes, which is why we extended those classes, as described in the previous paragraphs.

OCL Restrictions

A model is an abstraction of a system. In particular in this work we model a mining process. It should be as accurate as possible. In order to find an unambiguous and consistent model, it is necessary to reformulate the restrictions to the developed profile.

The language proposed by UML to specify the restrictions of the diagrams is the so-called OCL (Object Constraint Language) [27]. OCL offers a convenient way of getting models with a high degree of consistency, because it is a formal language used to describe keywords in the model [34], making them more accurate and less ambiguous.

OCL is fully integrated into UML. We can check the values of the model elements and set restrictions on them. An example of a restriction on the profile element, in this case the stereotype «Decide», is especify in Algorithm 1.

> Context Decide inv:
>
> ((self.incoming.source->for All(a | a.oclIs Type Of (Decide)) or
>
> (a.oclIs Type Of(Process))or (a.oclIsTypeOf (Divide)) or
>
> (a.oclIsTypeOf(Group)))and
>
> (self.outgoing.target->forAll(a | a.oclIsTypeOf(Decide)) or
>
> (a.oclIsTypeOf(Dispose))or (a.oclIsTypeOf(Process)) or
>
> (a.oclIsTypeOf(Divide))or (a.oclIsTypeOf(Group))))

Algorithm 1:

The restrictions fulfill the purpose of validating the model where the proposed stereotypes apply. If an error occurs, it is immediately observed. In this case, we can introduce corrections so that the model complies with the preestablished conditions and reconcile with the expected results.

CONCLUSIONS

UML allows expressing specific concepts of discrete event simulation in the field of mining, through the extension of its basic classes. In particular, simulation software modules like CREATE and PROCESS are easily identified with the elements of the activity diagram. This feature provides high functionality to the proposed extended model, because its behavior is similar to the simulated system. System validation which was obtained by applying the profile to the mining process, where the load-transport cycle is a key part of a reservoir, satisfied the requirements imposed by the daily production.

Having a standard language like UML implies that it is possible to obtain a model with stereotypes with well-formed rules without ambiguous concepts. In this paper we presented an extension of UML for the basic elements of the Arena simulation software, which provides a higher level of abstraction. The integrated vision of these two languages (UML and simulation) helps build a consistent model for the process being simulated. Therefore, combining both tools facilitates the development of a system.

The proposed profile presents advantages at two levels.

Metamodel Level. At this level, any UML tool supporting profiles can be used to import the mining profile defined in this work. In this way, any engineer can use any UML tool for building their mining models using the same modeling language (UML profile). The second advantage is that if using a simulation language, other than Arena, or if the same Arena is extended with new characteristics, those characteristics can be easily included into the profile. The engineers use a modeling language (UML profile) independent of the simulation language (Arena, Simula, GPSS, etc.). The mining UML profile can be included with the specific semantic of any simulation language. Finally, using formal languages supporting mathematic specifications such as OCL (first order logic, theory of set and theory of bag) ensures much more accurate models than the ones used by other traditional modeling languages such as ERD (Entity Relationship Diagram), DFD (Data Flow Diagram) as general modeling language or ExtendSim Suite, Arena Blocks as specific modeling languages. The mathematic specifications are incorporated as OCL specifications in the metamodeling level of the profile. Therefore, the proposed UML profile can be reused with any UML tool that includes profile definitions with OCL.

Model Level. At this level, the engineer builds its standard models using the proposed profile, and these models can be used in any mining environment. The stereotypes can be applied with any mining process simulation software (not only Arena). Moreover, any UML tool that allows incorporating the standard profiles according to the OMG can be used to modify these models. Finally, we obtained models verified mathematically with OCL. Those are more accurate models according to the restrictions that are specified in the profile level.

REFERENCES

1.   OMG Unified Modeling Languaje (OMG UML), Infrastructure, Version2.3, OMG Document Number: formal/2010-05-03, http://www.omg.org/spec/UML/2.3/.

2.   A. Watson, Visual Modeling: past, present and future, White paper UML Resource Page, 2008.

3.   J. Miller and J. Mukerji, "MDA guide," Tech. Rep., Object Management Group, 2003,http://www.omg.org/mda.

4.   G. Engels, J. H. Hausmann, R. Heckel, and S. Sauer, "Dynamic meta modeling: a graphical approach to the operational semantics of behavioural diagrams in UML," in Proceedings of the Unified Modeling Language (UML '00), vol. 1939 of Lecture Notes in Computer Science, pp. 323–337, 2000.

5.   "OMG Unified Modeling Languaje (OMG UML)," Superestructure, Version2.1.2, OMG Document Number: formal/2007-11-02, Standard Document, 2007,http://www.omg.org/spec/UML/2.1.2/Superstructure/PDF.

6.   J. Banks, J. S. Carson, and B. L. Nelson, Discrete-Event System Simulation, Prentice Hall, Upper Saddle River, NJ, USA, 2nd edition, 1996.

7.   SimPy, Copyright 2002–2012 Team SimPy, http://simpy.readthedocs.org/en/latest/.

8.   The Facsimile Simulation Library, Facsimile copyright 2004–2013, M. J. Allen, http://facsim.org/.

9.   Elmer-CSC". CSC-IT Center for Science Ltd. Retrieved 2010-06-24.

10.   W. D. Kelton, R. P. Sadoswski, and D. T. Sturrock, Simulación Con Software Arena, Cuarta Edición, McGraw-Hill, 2008.

11.   M. N. Nashalji, M. A. Shoorehdeli, and M. Teshnehlab, "Fault detection of the Tennessee Eastman process using improved PCA and neural classifier," Advances in Intelligent and Soft Computing, vol. 75, pp. 41–50, 2010.

12.   S. Yin, S. X. Ding, A. Haghani, H. Hao, and P. Zhang, "A comparison study of basic data-driven fault diagnosis and process monitoring methods on the benchmark Tennessee Eastman process," Journal of Process Control, vol. 22, no. 9, pp. 1567–1581, 2012.

13.   S. Yin, H. Luo, and S. X. Ding, "Real-time implementation of fault-tolerant control systems with performance optimization," IEEE Transactions on Industrial Electronics, vol. 61, no. 5, pp. 2402–2411, 2013.

14.   A. Teilans, A. Kleins, Y. Merkuryev, and A. Grinbergs, "Design of UML models and their simulation using Arena," WSEAS Transactions on Computer Research, vol. 3, no. 1, pp. 67–73, 2008.

15.   J. Barjis and B. Shishkov, "UML based business systems modeling and simulation," in Proceedings of the 4th International EUROSIM Congress, June 2001.

16.   O. Barbarisi and C. Del Vecchio, "UML simulation model for hybrid manufacturing systems," inProceedings of the13th Mediterranean Conference on Control and Automation, Limassol, Cyprus, June 2005.

17.   H. J. Pels and J. Goossenaerts, "A conceptual modeling technique for discrete event simulation of operational processes," IFIP International Federation for Information Processing, vol. 246, pp. 305–312, 2007.

18.   L. Fuentes-Fernández and A. Vallecillo-Moreno, "An introduction to UML profiles," The European Journal for the Informatics Professional, vol. 2, no. 2, pp. 6–13, 2004.

19.   Object Management Group, "Common warehouse metamodel (CWM) specification," OMG Document ad/2001-02-01, 2001.

20.   N. C. Debnath, A. Garis, D. Riesco, and G. Montejano, "Defining patterns using UML profiles," inProceedings of the IEEE International Conference on Computer Systems and Applications, pp. 1147–1150, March 2006.

21.   N. Debnath, C. A. Martinez, F. Zorzan, D. Riesco, and G. Montejano, "Transformation of business process models BPMN 2.0 into components of the Java business platform," in Proceedings of the 10th IEEE International Conference on, Industrial Informatics (INDIN), pp. 1035–1040, July 2012.

22.   N. C. Debnath, L. Baigorria, D. Riesco, and G. Montejano, "Metrics applied to aspect oriented design using UML profiles," in Proceedings of the 13th IEEE Symposium on Computers and Communications (ISCC '08), pp. 654–657, July 2008.

23.   Matlab, 1994–2013 The MathWorks, Inc, http://www.mathworks.com/products/matlab/.

24.   Y. Merkuryev, Riga Technical University. The Modelling and Simulation of Complex Systems: Methodology and Practice. An Overview. Information Technology and Management Science, 2012.

25.   R. Castro, E. Kofman, and G. Wainer, "A formal framework for stochastic discrete event system specification modeling and simulation," Simulation, vol. 86, no. 10, pp. 587–611, 2010

26. A. Kleins, Y. Merkuryev, A. Teilans, and M. Filonix, "A meta-model based approach to UML modelling and simulation," in Proceedings of the 7th WSEAS International Conference on System Science and Simulation in Engineering (ICOSSSE '08), 2008.

27. OMG Object Constraint Languaje (OCL), OMG Document Number: formal/2012-01-01. Standard document, http://www.omg.org/spec/OCL/2.3.1.

28. M. Bustillo Revuelta and C. López Jimeno, Recursos Minerales. Tipología, Prospección, Evaluación, Explotación, Mineralurgia, Impacto Ambiental. Madrid, 1996.

29. M. Bustillo Revuelta and C. López Jimeno, Manual de Evaluación y Diseño de Explotaciones Mineras. Madrid, 1997.

30. M. S. Abdullah, R. Paige, C. Kimble, and I. Benest, "A UML profile for knowledge-based systems modelling," in Proceedings of the IEEE 5th ACIS International Conference on Software Engineering Research, Management, and Applications (SERA '07), pp. 871–878, August 2007

31. C. Bartolini, A. Bertolino, G. De Angelis, and G. Lipari, "A UML profile and a methodology for real-time systems design," in Proceedings of the IEEE 32nd Euromicro Conference on Software Engineering and Advanced Applications (SEAA '06), pp. 108–115, September 2006

32. D. S. Frankel, Model Driven Arhcitecture. Applying MDA to Enterprise Computing, 2003.

33. IBM Rational Software Architect version 8.5. Copyright IBM Corporation 1987, 2012, http://www-01.ibm.com/software/rational/.

34. UML 2.0 OCL Specification, 2006, http://www.omg.org.

Chapter 2

Management of Sulfide-Bearing Waste, a Challenge for the Mining Industry

Björn Öhlander[1], Terrence Chatwin[2], and Lena Alakangas[1]

[1]Department of Geosciences, Luleå University of Technology, SE-971 87 Luleå, Sweden

[2]International Network for Acid Prevention, 2105 Oneida Street, Salt Lake City, UT 84109, USA

ABSTRACT

Oxidation of iron sulfides in waste rock dumps and tailings deposits may result in formation of acid rock drainage (ARD), which often is a challenging problem at mine sites. Therefore, integrating an ARD management plan into the actual mine operations in the early phases of exploration, continuing through the mine life until final closure might be successful and decrease the environmental impact. A thorough characterization of ore and waste should be performed at an early stage. A detailed knowledge of mineralogical composition, chemical composition and physical properties such as grain size, porosity and hydraulic conductivity of the different waste types is necessary for reliable predictions of ARD formation and efficiency of mitigation measures. Different approaches to prevent and mitigate ARD are discussed. Another key element of successfully planning to prevent ARD and to close a mining operation sustainably is to engage the mine stakeholders (regulators, community and government leaders, *non-governmental organization* (NGOs) and lenders) in helping develop and implement the ARD management plan.

INTRODUCTION

There has been a tremendous development of reducing the environmental footprint of mining the last several decades, but mining operations may still have detrimental effects on soil, water and biota. Mining operations generally require large areas of land, and associated conflicts arise that are primarily related to competing land uses. The mining industry is also a major energy consumer. In addition, a substantial amount of fossil fuel is used. Leakage of the nutrient nitrogen from undetonated explosives and from cyanide leaching for gold extraction is common. Dust and noise problems are common at mine sites. However, these effects occur only as long as a mine is active. The major potential long-term environmental effect of mining is formation of acid rock drainage (ARD) in sulfide-bearing mine waste, which can last for hundreds or even thousands of years (e.g., [1,2]). Volcanogenic massive base metal deposits contain a few percent of the valuable metals [3], and thus more than 90% of the ore will be waste after processing. Large porphyry copper ores often have an average copper concentration of less than 1%, resulting in that more than 99% of the ore will be waste after processing [4]. Gold is mined in deposits with a grade as low as a few grams per tons [5]. These examples show that the major parts of ores thus will be waste. In some mines, large amounts of wall rock have to be mined to get access to the ore, which results in waste rock deposits.

The Swedish biologist Carl von Linné observed early in the 18th century that the Falu River in south central Sweden was polluted by drainage waters from the Falu copper mine. However, it was only a few decades ago that ARD was recognized as a cause of serious damage to the environment. During the last 30 years, considerable effort has been invested worldwide by researchers, mining companies and environmental authorities to understand the fundamental processes occurring in weathering deposits of mine waste, and to develop cost-efficient technologies to prevent and control ARD [1, 2].

ARD may be formed in waste deposits containing Fe-sulfides such as pyrite (FeS_2) and pyrrhotite ($Fe_{1-x}S$, where $x = 0$–0.125), if the acid producing capacity of the Fe-sulfides is larger than the buffering capacity of carbonates in the waste (e.g., [1]). This ARD is often rich in heavy metals and metalloids. Conventional mining activities, in-situ mining and bio-mining not considered, generate two main types of wastes, which both may contain sulfide minerals. These waste types are waste rock (dominated by coarse material) that is removed to reach the ore, and finely ground tailings generated during the ore processing. When waste that contains Fe-sulfides is exposed to oxygen and moisture, these sulfides are oxidized to free Fe^{2+} ions and sulfate with the concurrent microbial catalyzed production of acidity. The oxidation rate depends on several factors including oxygen availability, temperature, pH, bacterial activity and surface area of pyrite grains [6, 7, 8]. This acidification enhances the mobility of heavy metals and other elements occurring in the mining wastes (e.g., [1, 6, 9, 10]. The free Fe^{2+} ions may further oxidize and precipitate as Fe-hydroxides generating further acidification, which often occurs in the recipient downstream. Waste from Cu, Zn, Pb, and Au mining usually contain Fe-sulfides, in contrast to waste from Fe-oxide mining. Oxidation of the Fe-sulfides in alkaline environments results in neutral drainage waters, with elevated concentrations of sulfate but most cations are secondarily retained by adsorption to mineral surfaces at pH > 6, while the concentration of anions may remain high [1].

The minerals industry is well aware of the potential adverse issues relating to Acid Rock Drainage. The costs carried by many mining companies to cover current liabilities are a clear indication of the gravity of the problem. To address these problems the International Network for Acid Prevention (INAP) has developed the GARD Guide, an internet-based best practices guide to prevent and mitigate the formation of acid, neutral and saline drainages from mining operations [2]. This guide was developed not only for mine operators, but for all minerals-industry stakeholders including regulators, financiers, communities and non-governmental organization (NGO's). In this paper, we will describe proven practices that prevent and mitigate ARD in mining and other excavations, mainly based on the GARD Guide [2], and the Swedish MiMi-programme [1].

RESULTS AND DISCUSSION-PREVENTION AND MITIGATION OF ARD

The primary approach to the prevention and mitigation of ARD is to minimize the supply of the primary reactants for sulfide oxidation, and/or maximize the amount and availability of acid-neutralizing reactants. These methods involve minimizing oxygen supply through decreasing oxygen diffusion or advection/convection, minimizing water infiltration and leaching (water acts as both a reactant and a transport mechanism), minimizing, removing, or isolating sulfide minerals, maximizing availability of acid neutralizing minerals and pore water alkalinity and controlling bacteria and biogeochemical processes. All of these actions control the pore water pH and its resulting metal leaching. In the case of water minimization, not only does it control the formation of ARD, but it also limits ARD and leached metals from being transported from the mine waste. Another prevention option is to remove the source, *i.e.*, Fe sulfides, from the mining wastes with the aim to reduce the total amount of ARD-producing waste and remediation efforts needed. Desulfurisation has been considered for some time as an alternative method to avoid formation of ARD [2]. The depyritized tailings must have so low sulfide content that they are not acid generating. Depyritised tailings could potentially be used as cover on other waste [2]. In the Minerals to Metal Initiative developed in South Africa [11] the desulfurization approach is developed in a systematic way. It is suggested that the main problem, the Fe sulfides, should not be spread out in tailings deposits as is the common practice today. The Fe sulfides should be concentrated, back-filled in mines, or deposited separately (e.g., [12]). Since pyrite may have relatively high concentrations of trace elements such as Au, Co and Cu [13, 14], and pyrite oxidation is a strongly exothermic reaction, another option is that metals and heat potentially could be extracted from these sulfides, today generally considered as problems and waste.

Mine waste needs to be managed using principles that control the environmental impact in both short and long term, as well as meeting the safety requirements on the deposits over long periods of time. While a certain impact on the local environment by disposal of significant amounts of mine wastes within the mine site is inevitable, the impact on the surroundings needs to be minimized. This waste minimization places high demands on mine operations management, and requires the use of appropriate waste disposal methods and ARD remediation.

A factor of particular importance is the need for methods that ensure safe disposal over very long periods of time. Neutralizing ARD by liming is common, but is a short-term solution that results in increased amounts of waste, although of another type [2]. Also other types of treatments of drainage waters from waste piles must be considered as short-term solutions. The environmental authorities and the mining industry in the Nordic countries for instance instead prefer remediation methods that last for very long times with a minimum of maintenance.

Prevention of ARD can be achieved through a risk-based planning and design approach that is applied throughout the mine-life cycle, but prevention is primarily planned and organized in the assessment and design phases and implemented during mine operations. The prevention process aims to quantify the long-term impacts of alternatives and to use this knowledge to select the option that has the least impact. Mitigation measures implemented as part of an effective control strategy should require minimal active intervention and management.

In general, more options and more effective options are available earlier in the mine life, as indicated in Figure 1 [2]. A thorough characterization of ore and waste should be performed at an early stage. A detailed knowledge of mineralogical composition, chemical composition and physical properties such as grain size, porosity and hydraulic conductivity of the different waste types is necessary for reliable predictions of ARD formation and efficiency of mitigation measures [2]. Different types of static and kinetic leaching tests are important in this context [2, 15]. More than one measure, or a combination of measures, may be required to achieve the desired objective. Sites that generate ARD with a high solute load and concentrations of contaminants can incur significant long-term ARD treatment costs that can impair the economic success and, in some cases, the viability of a project. Measures for ARD prevention, mitigation, and treatment must therefore be included in evaluation of mine lifecycle costs. The result of this overall assessment may be a decision not to mine a particular rock mass at some mines, or to mine in a manner that might initially be thought to be more costly [16].

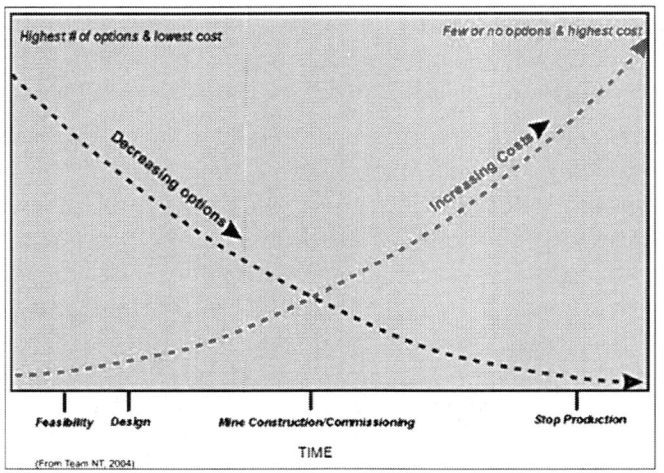

Highest # of options & lowest cost

Few or no options & highest cost

Decreasing options

Increasing Costs

Feasibility   Design   Mine Construction/Commissioning   Stop Production

TIME

(From Team NT, 2004)

Figure 1: Options and Effectiveness with Time (TEAM NT, 2004) [2].

Early avoidance of ARD problems is a best practice technique that may be achieved through integrating the results of characterization and prediction with mine planning, design and waste management strategies.

ARD Prevention and Mitigation Best Practices

A summary of the methods available for prevention and mitigation of mine drainage is shown in Figure 2. Detailed design manuals that were used in the compilation of these methods, such as [1,2,17,18,19,20] are listed in the references. All methods are site specific and need robust and flexible ARD prevention planning based on characterization and ARD prediction of the wastes in the early stage of the mine life.

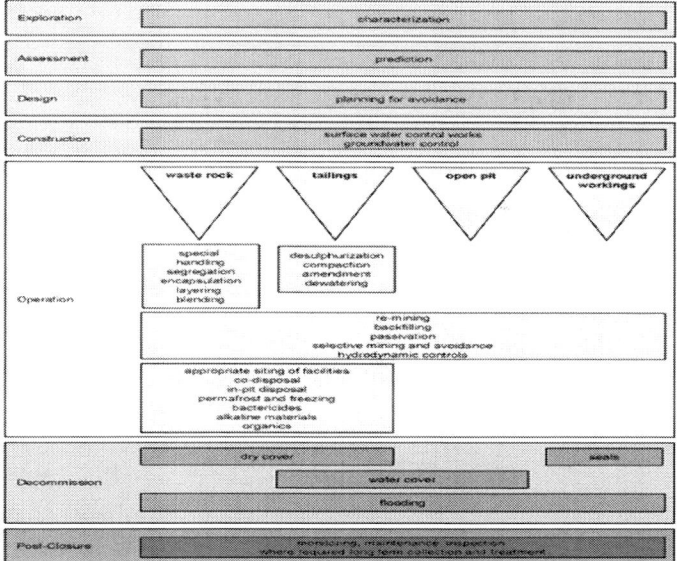

Figure 2: Methods for Prevention and Mitigation of acid rock drainage (ARD) [2].

During operation there is a need for reducing the environmental impact until the final remediation during mine closure. Proven preventative best practices include avoidance of acid producing minerals, re-mining of sulfide- and metal-rich wastes, special handling and segregation of sulfide-rich wastes. Mine-waste handling may be incorporated into mine planning to minimize exposure of materials to atmospheric conditions and minimize the volume of material left on the surface at closure. Examples of common practices used in integrating the ARD management plan into mine operations include the following:

- minimize the volume of sulfide waste by selective mining or sulfide-removal process steps so it can be adequately and cost effectively isolated from the environment.
- Encapsulation of the reactive tailings
- Mixing of tailings and reactive waste rock to minimize the oxygen ingress into the waste rock. This has been shown to be particularly effective when the tailings are dried to a paste form and mixed with the waste rock in cells or pits.

- Avoidance of placing waste storage facilities near sensitive receiving environments or regionally significant aquifers or recharge areas.
- Isolate waste repositories from uncontaminated surface waters through diversion ditches and dams.
- Isolate or segregate highly sulfide-rich tailings or mine waste using alkaline materials, covers, sub-aqueous disposal.

At mine closure, additional remediation efforts are often required such as water cover or dry covers. Different types of dry covers are often used to reduce the amount of oxygen reaching the waste. The most common methods are to apply dry covers consisting of several layers, usually including various soil types [1,2,17,18,19]. Both dry cover and water cover are based on the fact that the solubility and diffusivity of oxygen are much lower in water than in air. Dry covers, therefore, in most cases contain a sealing layer with low hydraulic conductivity, which is aimed at having a high degree of water saturation. The sealing layer then functions as a barrier against oxygen and water intrusion also in the cases when the groundwater surface is far below the cover. A protective layer is usually applied above the sealing layer to resist root penetration, freeze/thaw effects, drying etc. A type of dry cover often used by the Nordic mining industry consists of a sealing layer of clayey till with low hydraulic conductivity, and on top of that a protection layer of unclassified till. Modeling based on extensive lab- and field studies showed that the oxygen flux through this type of dry cover will be about one mole O_2/m^2 per year, a very strong reduction compared to pre-remediation conditions [1]. Also other materials such as cement-stabilized fly ash and organic waste (paper mill sludge and sewage sludge) have been used as sealing layers [2]. Sewage sludge has commonly been used for establishment of vegetation, but it has not often been used as a conventional barrier. Subaqueous disposal of reactive wastes in mine voids may be efficient, including placing mining wastes into open pits and underground workings. This method is particularly effective when the workings are flooded as the ground water recovers. The economic feasibility of this practice is highly site specific, but is fairly common, and the approaches are well developed. Mined-out pits can provide a void for storage of tailings, waste rock, or seepage water. Pits provide the potential for long-term geologically stable containment while traditional impoundments often require monitoring and maintenance to ensure stability of the constructed dam walls over the long term. In-pit disposal of tailings or waste rock may be combined with other strategies, such as subaqueous or underwater disposal, alkaline addition, cover technologies, and sulfate reduction. Water covers are often constructed by raising the dikes of tailings impoundments. However, this cannot be considered as a "walk away solution" without long time scale maintenance or supervision of the earthen dams, as demanded by the Swedish Environmental Protection Agency for instance with their "next ice age perspective". When functioning as planned, both dry and water covers slow down sulfide oxidation to an acceptable rate, resulting in that oxidation proceeds for very long times. However, there are risks for failures causing increased oxidation rates. According to the current policy of the Swedish Environmental Protection Agency, as an example, it is extremely difficult to get a permit to use natural lakes for disposal of mine waste, which otherwise in many aspects is the best solution.

The Figure 3 illustrates adaptive management and implementation using a phased approach, which begins with the development of hypotheses and conceptual designs based on site characterization and problem definition. The phased process can begin and enter at any stage of mine development. The key step is to develop a system design that leads to the basis for analysis and the capacity to make decisions. The process should include a staged approach that allows ongoing analysis, verification, and improvement in system design. Regional and local experience at nearby mines, where available, should also be used to minimize redundant investigations and to optimize the most successful methods for prevention and control. The phased approach leads to development of a monitoring and maintenance program that reinforces and improves system design.

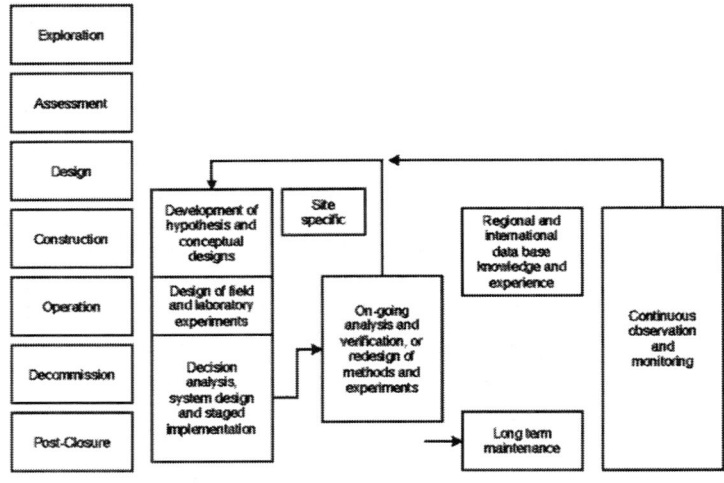

Figure 3: Adaptive Management Implementation by Phased Approach (from [2]).

Engaging Stakeholders

The proceeding sections addressed many of the technical aspects of ARD prevention and mitigation, but for this ARD prevention effort to be truly successful all of the stakeholders must "buy in" on the strategy and approach. To achieve this agreement, the mine stakeholders (regulators, community leaders, NGOs, lenders, governmental agencies, as well as the mine operators) must be engaged early in the decision process. Their expectations and perceptions must be heard and addressed so that they feel the ARD management plan contains much of their thoughts and decisions. At many mine sites there is a small, vocal group of antagonist that do not want a mine regardless of the value to the majority in taxes, jobs and economic well-being of the community. They only way that these antagonists can be answered is to be totally transparent and honest in the engagement process of listening, acknowledging, consulting, accepting and partnering with the stakeholders to produce an acceptable and effective ARD management and mine closure plan. The stakeholder engagement process should start early, involve key company management and be totally transparent for it to be successful.

Historical Mine Waste

Although metal emissions from mines under operation are generally low, leaching of metals from existing mining residues is a large problem. Effluents from historical mining represent an important environmental problem [5]. Strategies for remediation of existing waste deposits may differ somewhat from plans for future waste deposits. Waste deposits without cover have been exposed for oxidation, resulting in that they contain weathering products, secondarily retained within the deposits. In addition, if such waste deposits are isolated from oxygen entrance, Fe^{3+} may function an oxidant [1].

Several remediation options are possible for existing, ARD-producing waste [2]. Strategies are site specific and should be based on thorough characterisation of mineralogy, chemical composition and physical properties as well as of local geology and hydrology. Re-mining is one option, resulting in metal recovery and potential possibilities to store the processed waste in controlled tailings impoundments. Accelerated leaching by controlled bioleaching is another possibility to recover metals and reduce the amount of reactive sulfides in the waste. Mixing the waste with acid neutralising agents may be a way to reduce the weathering rate. These neutralising agents could be alkaline rest products from other industries, such as steel and paper industries. Soil cover or water cover may be applied also in the case of old waste deposits, but it is important to consider secondary effects such as wash out of old weathering products and of Fe^{3+} as oxidant [1, 2]. During a limited time period, it may therefore be necessary to combine the covers with active treatment of drainage waters. Typical treatments are to neutralise the drainage by using lime or alkaline rest products from other industries, or to use permeable, reactive barriers [21, 22].

CONCLUSIONS

ARD is a significant and potentially enduring environmental problem of the mining industry that could potently be the industries most lasting and harmful legacy. Therefore, integrating an ARD management plan into the actual mine operations in the early phases of exploration, continuing through the mine life until final closure might be succesful and decrease the environmental impact. Hence, it needs the commitment of the full mine operations team from mine planners to the mine manager, and it must be fully integrated into the mine economic model as well as operations early in the mine-life cycle. It should not be implemented as an after thought at closure when much of the human, financial and mechanical resourses are gone. A thorough characterization of ore and waste should be performed at an early stage. A detailed knowledge of mineralogical composition, chemical composition and physical properties such as grain size, porosity and hydraulic conductivity of the different waste types is necessary for reliable predictions of ARD formation and efficiency of mitigation measures. Best-practices methods for the prevention and mitigation of ARD should always be used. Another key element of successfully planning to prevent ARD and to close a mining operation is to engage the mine stakeholders (regulators, community and government leaders, NGOs and lenders) in helping develop and implement the ARD management plan. This engagement effort recognizes stakeholder concerns and perceptions and creates a partnership between the mine operator and the mine stakeholders in both the operation and closure of the mine.

ACKNOWLEDGEMENTS

In this paper we have mainly used material from the GARD Guide, and the authors wish to thank INAP member companies for their support to producing the GARD Guide. We have also used results from the Swedish MiMi-programme, financed by MISTRA.

REFERENCES

1.   Öhlander, B.; Neretniks, I.; Moreno, L.; Malmström, M.; Elander, P.; Lindvall, M.; Lindström, B.; Lövgren, L.; Herbert, R.; Höglund, L.-O. *MiMi—Performance Assessment, Main Report; MiMi-Report 2003:3*; Moreno, Italy, 2004.

2.   The global acid rock drainage guide (GARD Guide). Development of the Global Acid Rock Drainage Guide; International Network for Acid Drainage (INAP): Richmond, BC, Canada, 2011. Available online: http://www.gardguide.com (accessed on 20 December 2011).

3.   Franklin, J.M.; Gibson, H.L.; Jonasson, I.R.; Galley, A.G. Volcanogenic massive sulfide deposits. In *Economic Geology, One Hundredth Anniversary Volume*; Hedenquist, W., Goldfarb, R.J., Thompson, J.F.H., Richards, J.P., Eds., Eds.; Society of Economic Geologists,INC.: Pretoria, South Africa, 2005.

4.   Proffet, J.M., Jr.; Dilles, J.H; Seedorf, E.; Einaudi, M.T., Jr.; Zurcher, L., Jr.; Stavast, W.J.A., Jr.; Johnson, D.A., Jr.; Barton, M.D., Jr. Porphyry deposits: Characteristics and origin of hypogene features. In *Economic Geology*; Hedenquist, J.W., Thompson, J.F.H., Goldfarb, R.J., Richards, J.P., Eds., Eds.; Volume 100, pp. 251–298Society of Economic Geologists,INC.: Littleton, CO, USA, 2005.

5.   Lottermoser, B. *Mine Wastes, Characterization, Treatment and Environmental Impacts*; Springer-Verlag: Berlin, Germany, 2010.

6.   Jurjovec, J.; Ptacek, C.J.; Blowes, D.W. Mill tailings: Hydrogeology and geochemistry. In *Environmental Aspects of Mine Wastes, Short Course Series Vol. 31,Mineralogical Association of Canada*; Eds., Ritchie, A.I.M., Blowes, D.W., Jambor, J.L., Eds.; Mineralogical Association of Canada: Nepean, ON, Canada, 2003.

7.   Nordstrom, D.K. Effects of microbiological and geochemical interactions in mine drainage. In *Environmental Aspects of Mine Wastes, Short Course Series Vol. 31,Mineralogical Association of Canada*; Eds., Ritchie, A.I.M., Blowes, D.W., Jambor, J.L., Eds.; Mineralogical Association of Canada: Nepean, ON, Canada, 2003.

8.   Ritchie, A.I.M. Oxidation and gas transport in piles of sulfidic material. In *Environmental Aspects of Mine Wastes,Short Course Series Vol. 31,Mineralogical Association of Canada*; Eds., Ritchie, A.I.M., Blowes, D.W., Jambor, J.L., Eds.; Mineralogical Association of Canada: Nepean, ON, Canada, 2003.

9.   Holmström, H.; Salmon, U.J.; Carlsson, E.; Petrov, P.; Öhlander, B. Geochemical investigations of sulphide-bearing tailings at Kristineberg, northern Sweden, a few years after remediation. *Sci. Total Environ.* 2001, *273*, 111–133.

10.   Ljungberg, J.; Öhlander, B. The geochemical dynamics of oxidizing mine tailings at Laver, northern Sweden. *J. Geochem. Explor.* 2001, *74*, 57–72.

11.   Minerals to Metal Initiative; University of Cape Town: Cape Town, South Africa, 2009. Available online: http://www.mineralstometals.uct.ac.za (accessed on 3 December 2011).

12.   Harrison, S.T.L.; Broadhurst, J.L.; Hesketh, A.H. Mitigating the generation of acid mine drainage from copper sulfide tailings impoundments in perpetuity: A case study for an integrated management strategy. *Miner. Eng.* 2010, *23*, 225–229.

13.   Fleischer, M. Minor elements in some sulfide minerals. *Econ. Geol.* 1955, 970–1024.

14.   Gilbert, S.; Meffre, S.; Maslennikov, V.; Hollit, C.; Danyushevsky, L.; Large, R.L. Gold and trace element zonation in pyrite using a laser imaging technique: Implications for the timing of gold in orogenic and Carlin-style sediment-hosted deposits. *Econ. Geol.* 2009, *104*, 635–668.

15.   Price, W.A. *Draft Guidelines and Recommended methods for the Prediction of Metal Leaching and Acid Rock Drainage at Minesites in British Columbia*; Reclamation Section, Energy and Minerals Division, Ministry of Employment and Investment: Smithers, BC, USA, 1997.

16.   *Acid Drainage Technology Initiative (ADTI)*; National Land Reclamation Centre, West Virginia University: Morgantown, WV, USA, 1998.

17.   Hogan, C.M.; Tremblay, C. *Prevention and Control. Manual 5.4.2d G.A*; Mine Environment Neutral Drainage Program (MEND), The Canada Center for Mineral and Energy Technology (CANMET): Ottawa, ON, Canada, 2001.

18.   Design, Construction and Performance Monitoring of Cover Systems for Waste Rock and Tailings; Mine Environment Neutral Drainage Program (MEND), Report 2.21.4; Natural Resources Canada, O'Kane Consultants Inc.: Saskatoon, SK, Canada, 2004.

19.   *Macro-Scale Cover Design and Performance Monitoring Manual 2007; Mine Environment Neutral Drainage Program (MEND), Report 2.21.5; Natural Resources Canada*; O'Kane Consultants Inc.: Saskatoon, SK, Canada, 2007.

20.   *Best Practice Guideline—H2: Pollution Prevention and Minimization of Impacts*; Department of Water and Forestry, Republic of South Africa (DWAF): Pretoria, South Africa, 2007.

21. Mayer, K.U.; Ptacek, C.J.; Blowes, D.W.; Benner, S.G. Rates of sulfide reduction and metal sulfide precipitation in a permeable reactive barrier. *Appl. Geochem.* 2002, *17*, 301–320.

22. Hedin, R.S.; Banwart, S.A.; Younger, P.L. *Mine Water: Hydrology, Pollution, Remediation*; Kluwer Academic Publishers: Dordrecht, The Netherlands, 2002.

Chapter 3

Continental-Scale Assessment of Provisioning Soil Functions in Europe

Gergely Tóth, Ciro Gardi, Katalin Bódis, **Éva** Ivits, Ece Aksoy, Arwyn Jones, Simon Jeffrey, Thorum Petursdottir, and Luca Montanarella

Joint Research Centre, Institute for Environment and Sustainability, Via Fermi 2749 TP280, Ispra 21027, Italy

ABSTRACT

Introduction

A framework is developed to link major soil functions to ecosystem services assessment. Provisioning soil functions—with primary linkages to ecosystem services—are evaluated on a continental scale in Europe.

Methods

We defined major provisioning soil functions combining the approaches proposed by the Millennium Ecosystem Assessment and the Thematic Strategy for Soil Protection of the European Union. Soil productivity was evaluated by three main land use types (cropland, grassland, forest) using a validated expert model called SoilProd. Models include soil, climate and topographic criteria. Raw material provision capacity of soils was assessed on the basis of (i) organic carbon content and (ii) availability of soil materials for construction.

Results

A coherent system of soil function-based ecosystem services was compiled, taking into account major soil functions. We also produced new data on soil-based provisioning ecosystem services, including productivity and raw material availability. The attempts to cover the main human activities requiring materials of soil origin and to map the locations where those materials are available on a continental scale provide new insight to this field of research.

Conclusions

Soil-based ecosystem services can be assessed by the evaluation of soil functions which play a role in the production of these services. Quantitative analysis and comparison of the spatial distribution of the investigated soil functions were performed.

While crop productivity showed a general trend to increase in a northward and westward direction, local soil quality in most regions—except in the Mediterranean—can compensate for climatic handicaps to a great extent.

Comparison of areas with potential for providing ecosystem services by individual soil functions highlights the complexity of decision-making for resource utilization but also the possibilities for optimization and more conscious management.

INTRODUCTION

Soil plays a crucial role in terrestrial ecosystems and in maintaining life on Earth. Its functions which support ecosystem services to humans are manifold and complex.

This paper proposes a framework for the evaluation of soil functions that play a role in ecosystem services on a continental scale across Europe and offers an account of the repertoire of major soil functions and functional capacities of soils. Soil functions and associated services are discussed in the context of the European Union's Thematic Strategy for Soil Protection and are related to the concepts of the Millennium Ecosystem Assessment.

The attempt to characterize soil functions and ecosystem services for the terrestrial area of the EU in a spatially explicit manner is based on new data on provisioning soil ecosystem services, including productivity and raw material availability. Quantitative comparisons between major climatic zones of Europe were made with regards to the capacity of the soil to carry out provisioning functions.

Comparing the potential of individual soil functions across the EU highlights the complexity of decision-making dilemmas for resources utilization but also underlines the possibilities for resource use optimization and conscious management.

Soil plays a fundamental role in terrestrial ecosystems as a three-dimensional body that performs a wide range of ecological functions as a part of the services provided by ecosystems (Hannam and Boer 2004). The multitude of complex interactions which occur within the soil, including between biotic and abiotic compartments, gives rise to numerous soil functions (Blum 2005) which support ecosystem services, often viewed in terms of the services they provide to humans (De Groot et al. 2002; Fisher et al. 2009 Peccol and Movia 2012).

Soil functions are general or specific capabilities of soil to support various agricultural, environmental, landscape, and urban applications. Specific soil functions are manifold and may be grouped according to the Thematic Strategy for Soil Protection of the European Union (TSSPEU; CEC 2006) as (1) biomass production, (2) storing, filtering, and transforming nutrients and water, (3) hosting the biodiversity pool, (4) acting as a platform for most human activities, (5) providing raw materials, (6) acting as a carbon pool, and (7) storing geological and archaeological heritage. The focus of our current assessment has been on these seven major soil functions, which reflects our current scientific and policy perspectives towards Europe's soil resources (CEC 2006). Although the criticism may be valid that a scientific evaluation of soil functions as defined by a policy document is biased by current policy-driven perception, rather than being based purely on theoretical grounds (with possible windows towards applicability), it is generally accepted (FAO1976) that, in addition to studying soils' natural phenomena, soil resources should also be evaluated from socioeconomic viewpoints. This viewpoint is also an aim of the Millennium Ecosystem Assessment (MEA; Reid et al. 2005), the first comprehensive attempt to characterize the complex interactions between the various functions of ecosystems from the viewpoint of their services to humans. Therefore, a soil assessment derived from the recommendations of the TSSPEU and within the framework of soil functions is well justified for continental-scale applications in Europe.

A number of other classification and evaluation schemes for ecosystem services have emerged in recent years (Robinson et al. 2013), including the TEEB Assessment (2010; The Economics of Ecosystems and Biodiversity), which proposed a modification of the MEA scheme. Recommendations to link soil functions and ecosystem services have also been developed for local- to national-scale uses (Haygarth and Ritz 2009).

This paper aims to provide a framework to link major soil functions to ecosystem services assessment for continental-scale applications based on the concepts of the MEA and the TSSPEU. A further aim of the current paper is to provide an assessment of soil functions for provisioning services.

METHODS

Concept of the Assessment

The capacity of soil to perform any of the seven identified functions and to support their associated ecosystem services depends on its physical, biological, and chemical attributes (i.e., "internal" characteristics), while the realization of the performance is conditioned by natural (e.g., slope steepness) and/or anthropogenic factors ("external" controls). In order to provide quantitative figures on the levels of soil functions which are valid in the spatial context of the diverse biophysical regions of the European Union, evaluation of soil functions has to be extended and combined with landform and climatic information. As soil quality—the capacity of soil to perform its functions—is a time-dependent dynamic phenomenon (Karlen et al. 2001; Larson and Pierce1991), detecting temporal changes in soil functioning capacities is only possible if soil attributes are monitored. Since a comprehensive soil monitoring network and derived spatial database are lacking in Europe, the ambition to provide dynamic information on soil functions is unrealistic. However, characterizing soil functioning capacities on the basis of existing databases and matching this information with available land use and climate data were deemed feasible.

Most soil functions are interdependent; one function is not performed in isolation from another. Yet individual soil functions can be characterized by their primary linkages to ecosystem services and thus to services they provide to society, as is necessary for ecosystem services evaluation. Table 1 provides a framework proposed for continental-scale assessments. In the MEA, the process of soil formation, along with nutrient cycling and photosynthesis, was classified as an essential supporting service that facilitated other ecosystem services (i.e., provisioning, regulating, and cultural). As this study is concerned with soil functions, rather than underlying soil processes, a scheme modified from the MEA has been adopted, where soil functions are linked to ecosystem services through specific direct and indirect services provided by them.

Table 1: Soil functions that play a role in ecosystem services either directly, through their primary benefits to humans, or indirectly, by contributing to ecosystem functioning that eventually leads to benefits for humans

Soil functions (CEC2006)	Type of linkage	Ecosystem services (MEA2003)			
		Supporting	Provisioning	Regulating	Cultural
Biomass production	Direct		Providing food, feed, wood, fiber, biofuels		
	Indirect	Supporting ecosystem functions through primary production		Regulating carbon sequestration	Contributing to traditions, spiritual inspiration
Storing, filtering and transforming nutrients and water	Direct			Regulating water and nutrient availability	
	Indirect	Supporting ecosystem functions through water and nutrient cycling			
Hosting biodiversity	Direct	Supporting ecosystem functioning by hosting biodiversity			
	Indirect		Providing pharmaceuticals and biochemicals through gene reserve	Regulating crop pollination, pest, and disease control through living organisms	Contributing to scientific discovery
Platform for human activities	Direct	Supporting human habitats			
	Indirect				
Source of raw materials	Direct		Providing minerals and soil organic matter		
	Indirect				
Carbon pool	Direct			Regulating atmospheric CO_2	
	Indirect	Supporting soil structure, nutrient capacity, etc.		Regulating soil system	

Storing geological and archaeological heritage	Direct					Cultural heritage values (natural sciences, history, and anthropology)
	Indirect	Geological heritage supporting the maintenance of ecosystem dynamic equilibrium				

Tóth et al.

Tóth et al. Ecological Processes 2013 2:32 doi:10.1186/2192-1709-2-32

Soil Functions That Play a Direct Role in Provisioning Ecosystem Services

Soil functions include material outputs from soil, which can be organic or inorganic in nature. Organic materials include living aboveground biomass and subsurface organic matter. Inorganic materials include all forms of mineral compounds in the soil.

Biomass production and raw material provision are the two soil functions with material end products which are directly used by humans. However, it needs to be stressed that biomass production is a renewable function of soil, while the provision of raw material in most cases is destructive for the soil body. This distinction needs to be considered when planning for sustainable soil use.

To evaluate the performance of the biomass production function, traditional biophysical land evaluation methods can be used. For the evaluation of the raw material provision functions, the definition of the criteria for soil-originated raw material needs to be established. Firstly, it is important to set distinct criteria for soil and non-soil materials. Most mineral raw materials used by humans are of geological origin and are generally excavated by mining operations. Although soil parent material is usually considered as part of the pedon (e.g., soil taxonomy), in the raw material provisioning soil service, only soil layers above the parent material are considered. This designation also corresponds to the soil characterization provided by the TSSP, according to which "soil is generally defined as the top layer of the Earth's crust, formed by mineral particles, organic matter, water, air and living organisms" (CEC 2006).

Based on the above considerations, the biomass production function and raw material availability in the soil layers above the parent material were assessed during the evaluation of provisioning soil ecosystem services.

Biomass production of a given soil depends on the geographic location (climatic, hydrological, and terrain conditions), the land use type, land management, and the vegetation under consideration. In a continental-scale assessment, a land use–specific evaluation should be performed for the main biomass-producing land use types, namely cropland, grassland, and forest.

For continental-scale evaluations of raw materials in soils we must limit ourselves to the issues apparent over larger areas. While acknowledging the significance of locally important soil characteristics (such as high quality kaolinite for ceramics; bentonite for metal casting, well drilling, or food additives; clay for various applications), this study has focused on two applications: (1) organic soil (for horticultural and other applications) and (2) soil materials in construction. These two fields of applications involve by far the largest amounts of excavated soils in Europe. It is worth noting that a regional or local soil quality assessment might consider other elements (and with different weights). It is also worthwhile to mention that the growing demand for the services provided by the genetic pool of soil biota (e.g., from pharmaceutical industry) might lead to extraction of some soil material locally. However, the magnitudes of these kinds of extractions involve minor areas on the local scale and are not significant on a regional scale, and as such do not significantly affect the capacity of soil to provide other ecosystem services.

Attention should be called to the major difference between renewable biomass production and destructive use of soil substances. Authors strongly stress the need for considering sustainable soil use with regards to provisioning soil functions.

Databases

Soil Data

The European Soil Database (ESDB; EC 2003) was used as the soil information source in this study. Specifically, the Soil Geographical Database of Eurasia (SGDBE) and the PedoTransfer Rules Database (PTRDB) components of the ESDB were utilized.

The SGDBE, at scale 1:1,000,000, and the PTRDB are parts of the European Soil Information System (Panagos 2006; van Liedekerke et al. 2004). The SGDBE consists of both a geometric dataset and a semantic dataset (set of attribute files) which link attribute values to the polygons of the geometrical dataset. The database contains a list of soil typological units (STU). Besides the higher level soil taxonomic classification units represented by a soil name, these units are described by variables (attributes) specifying the nature and properties of the soils, for example, texture, water regime, or stoniness. As the original taxonomic information of the SGDBE is based on the 1990 FAO-UNESCO Soil Revised Legend (FAO 1990), these names and associated information content are used in the current exercise.

The PTRDB consists of a set of rules to derive new attributes with input attributes of the SGDBE (EC2003). Pedotransfer rules related to water availability were applied in this study.

Land Use Data

The land use data of the SGDBE were exploited during the biomass productivity evaluation. The SGDBE holds information on dominant and secondary land use types of each component STU of the soil map polygons. For cropland productivity evaluation, only those STUs were considered which had cultivated land as primary or secondary land use type in the SGDBE.

The GLOBCOVER dataset developed by the European Space Agency (Bicheron et al. 2006) was used to restrict the study area separately for the main land cover classes during the validation process.

The CORINE (CO-oRdination of INformation on the Environment; JRC-EEA 2005) land cover database for the year 2000 was used to select the extent of different land use types (cropland, grassland, forest land) to present the results.

Climate Data

Europe has diverse climatic conditions, represented by climatic zones and areas (FAO 1990; Köppen 1936). Rainfall characteristics and temperature regime seem to provide sufficient information for continental-scale agro-meteorological zonalization (Bouma 2005). The limiting effect of solar radiation is less articulated in most of Europe than that of rainfall or temperature. Differences in the radiation intensity can therefore be expressed through the above two factors and through terrain characteristics (slope and its orientation).

Soil quality evaluation for biomass productivity was therefore performed in a spatially explicit manner, taking the impact of the climatic component of soil productivity into account. Climatic zonation based on the 35 climatic areas of Hartwich et al. (2005) served as spatial units for soil productivity assessments on the continental scale. Regrouping of the climatic areas was performed to create climatic zones for soil quality assessment (Figure 1).

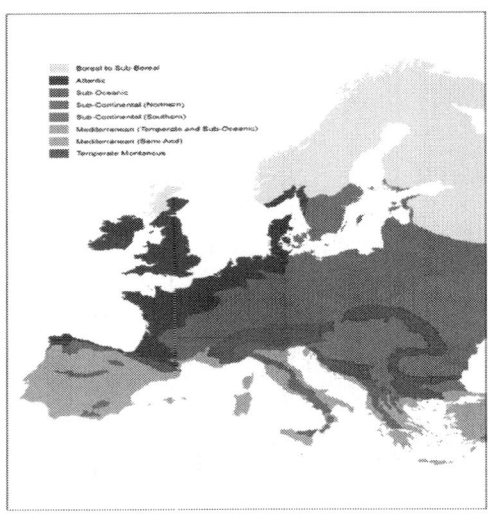

Figure 1: Climate zones of the study.

Crop and forest stand requirements regarding climatic conditions might differ to a great extent. For forest land, the length of the vegetative season and the soil water balance, calculated based on potential evapotranspiration (ETO; Penman 1948), were the aspects taken into consideration for the evaluation of forest productivity.

Topographic Data

A digital elevation model (DEM) was used to include a topographic component in the land evaluation model. The applied DEM was derived from the Shuttle Radar Topography Mission (SRTM; Rabus et al. 2003), which provided a dataset with a grid cell of 90 m.

Validation Datasets

In order to address the scientific reliability of the developed biomass productivity models, a validation procedure was applied using an independent dataset. A remote sensing–derived productivity indicator was used to validate all three land use–specific soil productivity models independently.

SPOT VEGETATION[a] decadal data were used to derive an approximation of biomass for the whole European continent. Data from the VEGETATION Program are frequently used in global and continental studies that supply input to General Circulation Models derived from measurements of the land cover and of the seasonal and long-term variations in vegetation dynamics. Studies that address the effect of biosphere processes and land cover characterization, the estimations of land cover variables as well as their dynamics, and the quantification of the mechanisms by which vegetation cover and ecosystems are interlinked all benefit from the VEGETATION Program. The SPOT data were corrected for system errors (misregistration of the different channels, calibration of all detectors along the line-array detectors for each spectral band) and resampled to geographic projections for multitemporal analysis as well as for comparison with high resolution data. For the present study, the 10-day synthesis or the maximum Normalized Difference Vegetation Index (NDVI) composite was used as is commonly accepted in the scientific community.

The NDVI time series was smoothed in order to remove short peaks and drop-offs due to noise. This pre-processing resulted in the reference time series from which the first (January) and last (December) absolute minimum values were derived for each year. The integral surface under the NDVI curve defined by the first and last minima of the year was derived and used as the approximation of net primary production (NPP; Figure 2). Ten years of NPP values were averaged for each pixel, and a final NPP dataset was produced (Figure 3). Many studies have already discussed the use of the integral surface of the annual NDVI time series curve as an approximation of aboveground biomass production (for an overview see Hill et al. 2008; Hellden and Tottrup2008). In the present study the NPP dataset was used to validate the soil productivity model in a geographically explicit manner.

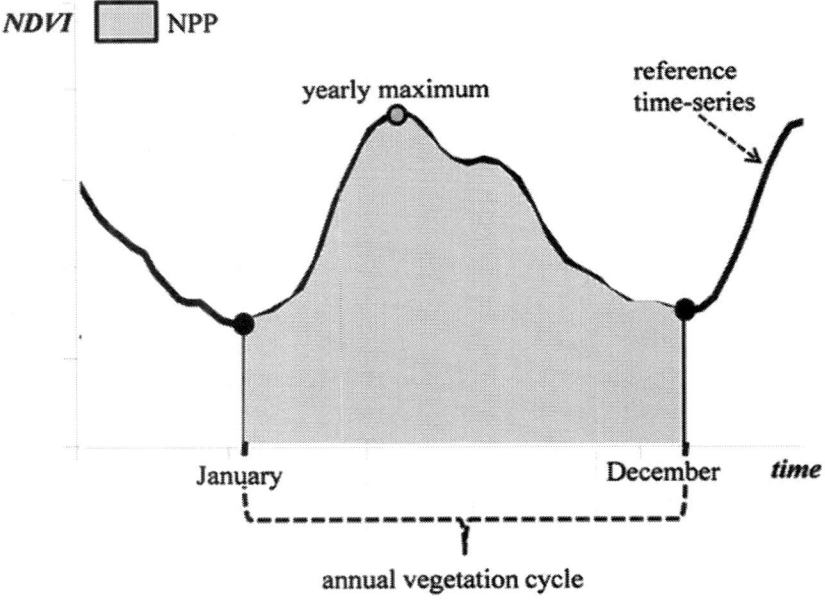

Figure 2: Schematic representation of gross annual biomass fraction (GABF) calculated from the NDVI time series for 1 year.

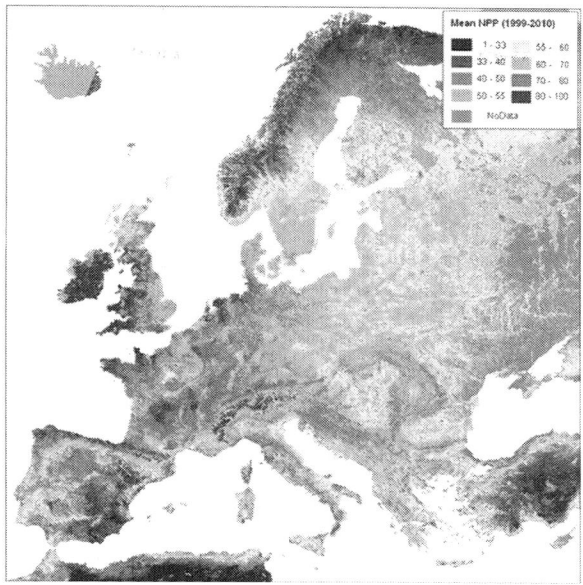

Figure 3: Mean GABF derived from time series of SPOT VEGETATION NDVI images averaged over the time series 1999–2010 for Europe.

For the raw material provisioning service, no validation datasets are available, therefore no similar validation was performed for this model.

Evaluation of Soil Provisioning Functional Capacities

Biomass Production

The degree to which the soil carries out its biomass production service was evaluated on the basis of soil properties under prevailing climatic and topographical conditions. Since productivity is a result of the interaction of soil, climatic, and topographical conditions, these factors need to be assessed in their complexity.

In addition to geophysical conditions, soil productivity also depends on the type of land use. The assessment of the European Environmental Agency (EEA 2006) shows that the three major land use types dominating the land cover of Europe are arable land with a share of 33%, pastures and mosaics with a share of 23%, and forests with a share of 29%. The aggregated share of these three types of land uses sums up to 85% of the total land and freshwater surfaces of the 24 countries of Europe assessed by the EEA (2006). Besides these major land use types, there are a number of specific regionally characterized land uses in Europe. There might also be considerable differences in the land utilization within the main land use types. However, for a continental-scale assessment of biomass productivity, the productivity patterns were evaluated according to the three major land use types. Models were therefore developed to describe general orders of soil productivity within the three land use types, namely for pasture/ grassland, cropland, and forest.

Calculations were performed in a spatially explicit manner, taking climatic and topographical conditions into account. Productivity models were built to reflect rain-fed conditions. The description of temporal variability of productivity or the estimation of provision productivity by means of actual yields was not among the aims. Results are presented in land use–specific maps (e.g., cropland productivity for areas of rain-fed arable lands, forest biomass productivity for forest lands, and grassland productivity for pastures and mosaics).

Biomass Production on Grasslands and Arable Lands

Productivity differences of similar soils under intensive rain-fed agricultural use and under grassland land cover vary with the changing availability of precipitation and differences in temperature regimes. For instance, in a temperate sub-oceanic climate, the rather stable thermal regime and balanced water availability of medium to high amounts of precipitation not only secure plant-available water on most soil types throughout the growing period but facilitate decomposition and weathering throughout the year. These processes are limited under temperate continental and Mediterranean climates due to cold and/or arid periods in most years. With increasing aridity in prevailing climates, the importance of soil physical and chemical properties in water and nutrient supply to plants gains increasing significance. Based on this principle of soil productivity processes, ranking of inherent productivity of soil was performed separately for the major climatic areas. While fertility is defined as the ability of the soil to provide nutrients and water, productivity refers to the capacity to supply nutrients and water and thus produce plant biomass at a given level/quantity. Inherent productivity in this context means soil productivity before human interference. In this study, inherent soil fertility—in this context productivity—is considered as a proxy for grassland productivity. Although the existence of different nutrient inputs on grasslands in different regions of the European Union is recognized, due to the lack of adequate data, grassland fertilization and other management factors were not taken into account. Cropland soil productivity was evaluated by the extension of inherent productivity with a management factor, as described below.

Grassland/Pasture

In the first step of the evaluation process, eight characteristic European climate systems were identified where the complex effects of water availability and thermal regime are distinct for soil processes and plant growth. The 35 climatic areas of Europe (Hartwich et al. 2005) were arranged into eight climatic groups accordingly (Figure 1). The climatic groups comprise regions where the following concepts of soil productivity processes prevail: boreal to sub-boreal (CZ1), Atlantic (CZ2), sub-oceanic (CZ3), northern sub-continental (CZ4), southern sub-continental, (CZ5), Mediterranean semi-arid (CZ6), Mediterranean (temperate and sub-oceanic) (CZ7), and temperate mountainous (CZ8). Long-term average inherent soil productivity ratings were assigned to soils of each region in a spatially explicit manner. Temporally variably climatic hazards of productivity (frost, water logging, etc.) were not taken into account in the current study.

Inherent soil productivity estimates were derived from the original taxonomic component (second-level taxonomic soil units) and soil attribute information of the soil database. Second-level taxonomic soil units were grouped into five inherent capability classes according to their relative productivity in each climatic zone. CZ1 includes 59 different soil units, CZ2 has 113, CZ3 has 122, CZ4 has 117, CZ5 has 128, CZ6 has 96, CZ7 has 103, and CZ8 has 131. This approach allows productivity assessment of soil types at different climatic regions. Table 2 provides an example of the classification.

Table 2: Examples of climate-zone-based productivity classification reference (10 soil units out of the total 128 present in the southern sub-continental climate zone are displayed here)

Inherent capability classes (southern sub-continental climate)				
V.	IV.	III.	II.	I.
Calcaric Lithosol	Orthic Rendzina	Gleyo-Dystric Luvisol	Albic Luvisol	Haplic Phaeozem
Gleyic Solonchak	Gleyic Podzoluvisol	Dystric Regosol	Chromic Cambisol	Luvic Chernozem

Tóth et al.

Tóth et al. Ecological Processes 2013 2:32 doi:10.1186/2192-1709-2-32

In parallel, STUs were also rated according to the available water capacities (AWC) of topsoil and subsoil. According to the pedotransfer rules of the PTRDB (EC 2003) and based on physical soil properties and soil depth, AWCs of both the topsoil and the subsoil were calculated. Soils were grouped into four classes on the basis of the water-storing capacity of the profile (Table 3).

Table 3: Examples of water-storing capacity classifications

Water-storing capacity classes	AWC of topsoil	AWC of subsoil	Depth to impermeable layer
I.	Very high	Very high	Deep
	High	Very high	Deep
II.	Medium	High	Deep
	High	Medium	Deep
III.	Medium	Low	Deep
	High	Very low	Deep
IV.	High	Very low	Shallow
	Medium	Low	Shallow

Tóth et al.

Tóth et al. Ecological Processes 2013 2:32 doi:10.1186/2192-1709-2-32

In the next step of the productivity assessment, an evaluation matrix was set up for the eight climatic regions and the five inherent capability classes. Productivity scores between 1 and 8 were assigned for each cell in the matrix, based on the complex evaluation of the climate-dependent relative inherent fertility of soils (Table 4).

Table 4: Productivity scores for the soil capability classes by climatic zones

CZ1	CZ2	CZ3	CZ4	CZ5	CZ6	CZ7	CZ8	Score
								8
I.	II.	I.	I.				I.	7
II.	III.	II.	II.		I.	I.	II.	6
III.	IV.	III.	III.	I.	II.	II.	III.	5
IV.	V.	IV.	IV.	II.	III.	III.	IV.	4
V.		V.	V.	III.	IV.	IV.	V.	3
				IV.	V.	V.		2
				V.				1

Tóth et al.

Tóth et al. Ecological Processes 2013 2:32 doi:10.1186/2192-1709-2-32

Water-storing capacity classes (Table 3) were assigned a multiplication factor between 0.75 and 1 for each climate zone. These factors were used to multiply the productivity scores of each soil unit of the climate zones.

The position of soil units in the capability classification matrix, productivity scores of capability classes by climate zones, and multiplication factors to account for the water-storing capacities by climate zones were first defined by expert judgment and then refined and completed through an iterative analytical process with a series of cross validations using regression analysis (GWR) with the validation dataset as dependent variable (see below for details on the validation process).

The result of this procedure is the soil productivity index for each soil typological unit in the climate zones. The index score 1 represents the poorest and 8 the highest productivity soil. The corresponding inherent productivity scores were assigned to each STU in the SGDBE. A spatially weighted average of inherent productivity score was calculated for each soil mapping unit on the basis of the proportional areal shares of the STUs within the mapping units.

Correction measures based on topographic conditions were finally applied to arrive at the final evaluation scores of inherent soil productivity. Correction coefficients of slope and aspects applied according to the D-e-Meter land evaluation system (Tóth 2009) were adapted for this continental-scale study.

Croplands

To evaluate the biomass productivity of soils under arable cultivation, the model was extended with a new module. Since the productivity of soil is due to its inherent fertility and to the effect of management—mainly nutrient input—the effect of fertilization was considered in this module. While acknowledging the importance of the applied technology of soil use on the actual productivity of soil, detailed distinctions of management practices were not considered in this study. The goal of this study was solely to determine soil productivity, i.e., the capacity of soil to supply nutrients, water, and rooting medium for plants, in a comparative manner and not to assess the effects of management differences other than nutrient supply. We also assume that in the case of the realization of biomass productive capacity, technological advancement plays a relatively insignificant role within distinct regions in European Union. However, in certain regions, where complex socioeconomic and biophysical conditions diversify the technological levels, this statement might not hold. The efficiency of input use, i.e., the selection of the most appropriate techniques and the amount of input, is largely determined by biophysical conditions (e.g., large regional crop yield differences within France (Eurostat 2013), are due to biophysical differences rather than differences in available technology, capital, or other socio-economic conditions). However, the technological advancement and input intensity of agriculture across regions in the European Union varies, and this variation contributes to differences in crop yields.

It was not the goal of this study to evaluate all management-related yield responses of soils, but rather, to evaluate the effect of management, we considered the influence of fertilization. To do this, second-level taxonomic soil units were grouped into five classes in each climatic zone, according to the magnitude of their expected productivity increase due to fertilization (Table 5).

Table 5: Sample reference table for classifying the response of different soil types to fertilization on a climate zone basis (10 of 128 soil units present in the southern sub-continental climate zone are shown here)

Fertilizer response classes (southern sub-continental climate)				
V.	IV.	III.	II.	I.
Luvic Ranker	Eutric Gleysol	Calcaric Phaeozem	Albic Luvisol	Dystric Fluvisol
Cambic Rendzina	Chromic Vertisol	Dystric Podzoluvisol	Calcic Cambisol	Calcaric Cambisol

Tóth et al.

Tóth et al. Ecological Processes 2013 2:32 doi:10.1186/2192-1709-2-32

Then, a fertilizer response score was assigned to each soil unit in the eight climatic zones. Soils with the largest relative fertility increase received the maximum of 8 points and soils with little influence of fertilization received 1 point. For example, an Albic Arenosol in the sub-oceanic climate has among the highest relative productivity increases when properly fertilized, while fertilization has little effect on the crop productivity of Calcaric Rendzinas in the Mediterranean.

To calculate soil productivity for the cropland land use type, the inherent soil productivity and the fertilizer response scores were aggregated, assigning a mechanical weight to the fertilizer response indices. This weight resulted in the best model fit at the end of an iterative statistical validation process (see below for details of the statistical validation). Spatially weighted averages of productivity scores were calculated for the SMUs of the SGDBE. In order to avoid the bias originating from the evaluation of non-cropland soils, only those STUs were considered which had cultivated land as the primary or secondary land use type in the SGDBE. Finally, similarly to the concluding step of the grassland productivity evaluation, correction coefficients were applied to evaluate the effect of the topography (slope and aspect) on the productivity of cropland soils.

Biomass Production on Forest Lands

The productivity of a forest, in terms of biomass, is the result of the interaction of tree species, soil, and climate. Ecosystem research has provided sufficient information about key processes such as photosynthesis, transpiration, and decomposition to allow the construction of simulation models that predict properties such as biomass production, soil carbon storage, and nitrogen cycling rates (e.g., Aber et al. 1997; Parton et al. 1988; Raich et al. 1991; Running and Gower1991).

Several of these models are based on complex input data sets, consisting of a series of climate, soil, and forest stand data (i.e., maximum and minimum daily temperature, vapor pressure, solar radiation, total monthly precipitation, forest type, and plant-available soil water holding capacity).

The aim of this study, however, was to predict, in general terms, the forest productivity potential of European soils. For this purpose, two main processes related to forest productivity were evaluated: the length of the vegetative season and the soil water balance. The concept behind this simplistic approach is that the net ecosystem productivity (NEP) of a forest can be limited by two main factors: the air temperature, which determines the length of the vegetative season, and the amount of water from soils that the plants can evapo-transpire. Accordingly the northern latitude forests will be limited by the relatively short vegetative season, while the Mediterranean forests will be limited by the relatively low water availability, which in turn is related to both climate and soil characteristics.

The equation used for the calculation of forest productivity is shown below:

$$\text{Forest productivity} = (SWBNx/SWBNmax) \times \text{vegetative length}$$

where SWBNx is the soil water balance normalized at a given location (mm), SWBNmax is the maximum soil water balance (mm), and vegetative length is the length of the vegetative season (days). The soil water balance is calculated as:

$$\Sigma(Pi-ETPi) + AWC \ (mm)$$

where Pi is monthly precipitation, ETPi is monthly potential evapotranspiration (Kc =1), and AWC is the available water content (mm).

The values obtained, ranging from 0 to 313, were grouped into 10 equal classes. The output map was masked based on the actual forest distribution as defined by the CORINE Land Cover database.

Results of all three productivity evaluations (grassland, cropland, forest) were scaled to index scores ranging from 1 to 10 showing the relative fertility of soils expressed in relative index values without units.

Provisioning Raw Materials

The evaluation of raw material availability from soil origin is limited to resources above the parent material. The assessment included the following two options, with the relevant criteria:

- Peat (for horticultural and other applications) and organic topsoil: All organic soils (Histosols) are considered and no mineral soils are considered.

- Soil materials for construction: To assess the quality of soils to provide construction materials, the approach presented in the Soil Atlas of Europe [b] was applied, and the presence of sand and gravel was examined for this function. Criteria:

- Coarse texture (clay < 18% and sand > 65%) and/or
- Stones and gravel content are dominant in the horizon (> 80% by volume)

In addition to the above construction materials, loamy clay is often used to produce bricks and tiles. Although the Soil Atlas of Europe does not consider these as among the dominant soil-based construction materials, their widespread usage could justify the inclusion of loamy-clay (sub) soils in the analysis. However, the currently available continental soil databases contain no information on soil texture at this level of detail; therefore such an analysis was not feasible at this time.

The evaluation did not account for any economic considerations (i.e., only the availability of raw material was assessed). Results are presented as the proportional availability of these materials in the mapping units. For Cyprus, no information on texture and stone and gravel content was available and, therefore Cyprus was excluded from this analysis.

Statistical Validation

The geographically weighted regression (GWR) method (Brunsdon et al. 1996; Fotheringham et al.2002) was selected to validate the model results against the measured productivity indicators. The GWR method was selected as it allows for the possibility of assessing relationships in spatially explicit information. Spatial data are rarely consistent due to random sampling variations, the different relationships between variables across space, or model misspecifications of reality (Fotheringham et al. 2002). However, the application of standard regression methods to spatial data would result in the assumption of the spatial phenomenon being constant over space.

GWR is a statistical technique that allows the modeling of processes that vary spatially. Consider the ordinary least squares (OLS) regression model given by

$$y_i = a_0 + \sum_k a_k x_{ik} + e_i,$$

where e_i is the error term, a represents the vector of global parameters to be estimated, x is a matrix of independent variables and y represents a vector of observations on the dependent variable.

GWR extends the traditional regression framework by allowing local rather than global parameters to be estimated. Thus the OLS regressions model is rewritten as:

$$y_i = a_0(u_i, v_i) + \sum_k a_k(u_i, v_i)x_{ik} + ei,$$

where (u_i, v_i) denotes the coordinates of the ith point in space and $a_k(u_i, v_i)$ is a realization of the continuous function $a_k(u, v)$ at point i (Brunsdon et al. 1996).

This extension allows measurements of a continuous surface of parameters to be taken and to denote the spatial variability of the surface because observed data near point i have more of an influence in the estimation than do data located farther from i. Thus GWR becomes a weighted least squares regression where observations are weighted in accordance with the proximity to point i and the weight decreases with increasing distance from with i.

The biomass productivity spatial datasets were divided—both from our land evaluation exercise and from the satellite observations—by land use, based on small regions for special diagnoses (NUTS3; Eurostat 2011). Analyses of statistical correlation between the datasets were performed using the GWR method as described above.

Per pixel biomass values estimated from the land evaluation exercise and from the satellite-based observations were averaged within the regions of the NUTS3 dataset (Eurostat 2011) for the countries of the European Union. The GWR analysis was carried out separately for croplands, grasslands and forests with the remote sensing–derived biomass estimates as the dependent variable and the land cover–specific biomass estimates as independent variables. Each regression run was weighted by the number of pixels within the NUTS3 polygons.

The result of the GWR model for croplands is valid with a goodness of fit characterized by a coefficient of determination (adjusted r^2) of 0.73. The adjusted r^2 figure for grassland is 0.85 and for forests 0.90. These results can be considered as a confirmation of the validity of the soil productivity evaluation approach.

RESULTS AND DISCUSSIONS

Grassland Soil Productivity Map of the EU

The application of a geographically explicit spatial soil grassland productivity model showed high consistency with the data obtained from remote sensing ($R^2 = 0.85$).

Based on the cumulative relative productivity indices, the soils in the Atlantic climatic zone have the highest potential for grassland production. They contribute nearly half of the potential for the European Union (44.6%), although only 38.4% of Europe's grassland areas are found in the Atlantic zone (Table 6). The zone productivity factor of grassland soils in this climate zone is 1.2, indicating 20% higher relative productivity indices on average when compared to the mean of the EU. In contrast, grasslands of the semi-arid Mediterranean region have relative productivity indices that are 40% lower than the average of the EU. Apart from the northern sub-continental zone

(zone productivity factor = 1), all other zones have relative productivity indices for grasslands that are lower than the average of the EU. This fact indicates that, in most of the climatic zones (CZ1, CZ3, CZ5, CZ6, CZ7, CZ8) of Europe, grassland ecosystems are on soils of lower production potentials.

Table 6: Summary statistics of productivity indices for soils in the European Union by major land use classes and main climatic zones

Climate zonea	Area		Productivity index				Zone productivity factorc
	km2	Percentage of EU	Mean	Standard dev	Sum	Percentage of totalb	
Grassland							
1	21,571	4.5	5.8	1.7	124,963	4.3	0.9
2	184,363	38.4	7.1	2.1	1,305,600	44.6	1.2
3	88,910	18.5	5.8	2.0	515,760	17.6	0.9
4	55,982	11.7	6.3	1.3	350,335	12.0	1.0
5	33,309	6.9	5.8	1.2	192,966	6.6	0.9
6	21,959	4.6	3.8	1.2	84,031	2.9	0.6
7	16,170	3.4	4.2	1.3	68,612	2.3	0.7
8	57,306	11.9	5.0	1.4	286,378	9.8	0.8
Cropland							
1	32,458	2.4	6.1	1.0	198,645	2.5	1.0
2	308,693	23.0	7.1	1.3	2,189,030	27.6	1.2
3	170,655	12.7	5.7	1.6	974,337	12.3	1.0
4	337,121	25.1	6.2	1.1	2,099,960	26.5	1.1
5	175,914	13.1	5.8	1.0	1,013,880	12.8	1.0
6	174,171	13.0	4.0	1.2	698,184	8.8	0.7
7	109,687	8.2	5.0	1.2	545,528	6.9	0.8
8	35,101	2.6	5.7	1.3	198,537	2.5	1.0
Forest							
1	531,907	30.9	2.3	0.8	1,202,780	19.2	0.6
2	120,702	7.0	5.0	1.7	606,879	9.7	1.4
3	233,015	13.5	4.8	1.2	1,111,580	17.8	1.3
4	276,591	16.1	3.3	0.8	923,848	14.8	0.9
5	86,110	5.0	4.5	1.1	385,547	6.2	1.2
6	113,045	6.6	4.1	1.1	460,481	7.4	1.1
7	151,472	8.8	4.8	1.5	731,081	11.7	1.3
8	207,675	12.1	4.0	1.2	836,053	13.4	1.1

[a]1: Boreal to sub-boreal, 2: Atlantic, 3: sub-oceanic, 4: sub-continental (northern), 5: sub-continental (southern), 6: Mediterranean (semi-arid), 7: Mediterranean (temperate and sub-oceanic), 8: temperate mountainous. bExpressed in terms of productivity index.

cZone productivity factor = % of total productivity indices/% of total area in the EU (h = g/b).

Tóth et al.

Tóth et al. Ecological Processes 2013 2:32 doi:10.1186/2192-1709-2-32

The dependency of productivity on climate was statistically significant at the continental scale. Up to an estimated average of 60% of the grassland productivity differences between the most favorable (Atlantic) and least favorable (Mediterranean semi-arid) regions of the EU can be attributed to climatic factors, as indicated by the differences in mean productivity indices between the climatic zones and standard deviations within the climatic zones (Table 6). The influence of regional soil variability expressed as the standard deviations of the soil productivity indices might exceed the figure attributed to the influence of climate between different climate zones. These results confirm the general understanding of the importance of water availability in soil and the synergism between optimum temperature, water, and nutrient regimes and also present these interdependencies in a quantified manner in a spatial context.

Results of the inherent soil productivity evaluation are shown in the grassland productivity map of the European Union (Figure 4).

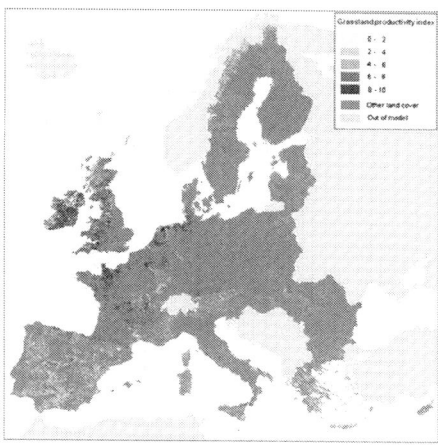

Figure 4: Soil biomass productivity of grasslands and pastures of the European Union.

Cropland soil productivity map of the EU.

A comparative analysis between modeled and remote sensing–derived soil productivity indicators demonstrated a significant climate dependency in the general pattern of cropland productivity of soils in the EU. However, a strong influence of soil type and soil properties was also verified.

Productivity showed a general trend of increasing in a northward and westward direction, however, local soil quality in most regions—except in the Mediterranean—can compensate for the climatic "handicap" to a great extent. While soils of the Atlantic region (CZ2), as was the case of the grassland ecosystem, have the highest relative potential to perform biomass provisioning ecosystem services in the cropland agro-ecosystem, this relative advantage is not as pronounced as in the case of grasslands (Table 6). In fact, soils of the northern sub-continental zone (CZ4; with 25.1% of the croplands of the EU) also perform above the EU average (zone productivity factor = 1.1), and another four zones (CZ1, CZ3, CZ5, CZ8), with a total share of 31% from the croplands of the EU, have, in general, soil productivity near the average of the continent (zone productivity factor = 1; Table 6). Among these four zones, the sub-oceanic zone (CZ3) possesses the highest intra-variability, as indicated by the standard deviation of the productivity assessment. Cropland soils of the Mediterranean (CZ7 and CZ8) with a combined area share of 21.2% have substantially lower rain-fed crop productivity compared to the EU's average, demonstrated by zone productivity factors of 0.8 and 0.7, respectively.

It is worth noting that irrigation can significantly increase the productivity of soils in the Mediterranean and to some extent also in the sub-continental region, where water is the limiting factor for crop growth. However, to assess (actual and potential) irrigated productivity of soil was out of the scope of this study. Nevertheless, future research should be carried out to analyze this aspect as well.

Results of soil productivity evaluation for croplands of the European Union are presented in Figure 5.

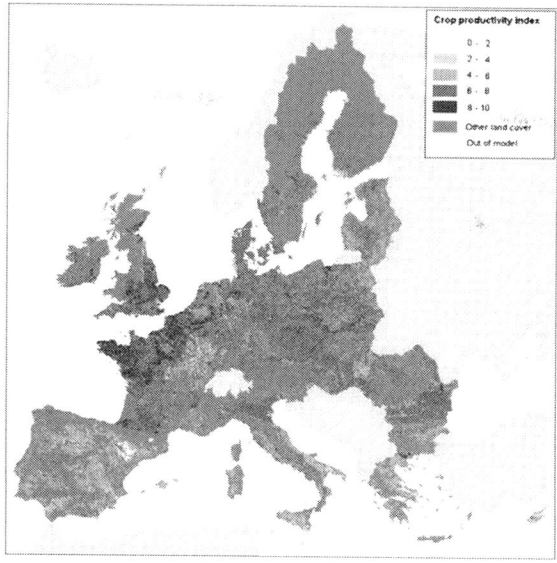

Figure 5: Soil biomass productivity of croplands in the European Union.

Forest Soil Productivity Map of the EU

Forest productivity is essentially the expression of the interaction between climate and soil factors. The productivity of forest soils is a function of their ability to hold and make water available to trees (available water capacity, AWC). Their nutrient status is also important. Available water capacity is a function of the rooting depth and texture of soils.

A common measure, especially in United States, for the evaluation of the collective influence of soil factors on forest growth is the site index (SI). The most important factors determining the SI are the topsoil depth and soil texture. The AWC parameter is strongly correlated with soil depth and soil texture and consequently can be used as a proxy of SI.

From the data shown in Figure 6, the climate driver is the most evident at larger scales, while the influence of soil can be detected mainly at more detailed scales. In fact, zone productivity factors (Table 6) have the widest range across the climate zones in Europe for forest land among the three land use classes assessed. The zone productivity factor is 1.4 for the Atlantic zone (CZ2), but only 0.6 for sub-boreal to boreal climates. In general terms, it is clear that climate is responsible for forest productivity variation at the regional scale while soil characteristics cause local (i.e., at scales of 10–100 km^2) variations.

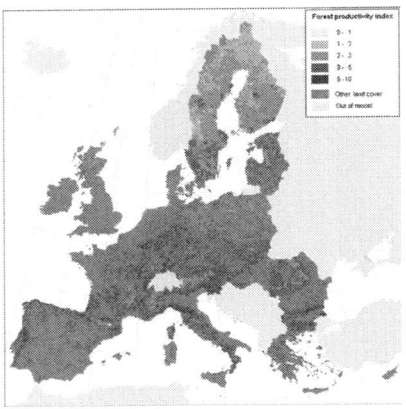

Figure 6: Soil biomass productivity of forest areas in the European Union.

It is evident that in the northern areas the favorable moisture conditions are negated to some extent by the shorter length of the vegetative season, while in the Mediterranean climate the length of the vegetative season is not a constraint and water availability during summer represents the most important limiting factor.

The soil parameter taken into consideration in the proposed approach was the AWC, for which only three classes are available within the European Soil Geographic Database: 120, 165, and 220 mm. The effect of soil is evident at a local scale, where it can result in a large variation in the absolute values of soil productivity.

Continental-Scale Map of Soil-Extractable Raw Materials of the EU

Raw material provisioning of the mapping units was calculated on the basis of the proportional shares of the STUs with raw material content in the area of each mapping unit. Once again it is important to stress that geological maps showing availability of materials worthwhile for human use that are present below the soil cover were not considered in this assessment and might show a very different pattern. The result of the assessment is presented in Figure 7.

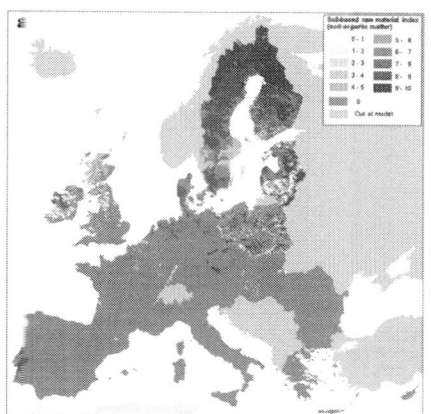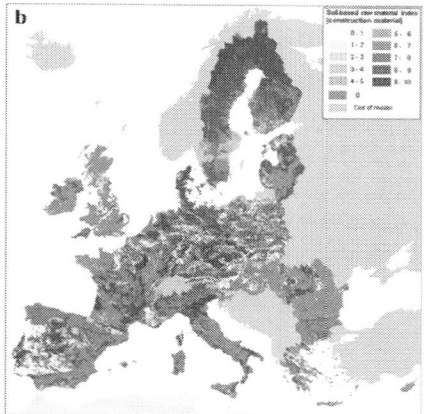

Figure 7: Raw material availability from soils of the European Union. a) organic soil material; b) soil material for constructions.

It is well acknowledged that this first approximation to highlight the availability of soil-born raw materials on the continental scale might be biased (1) by the classification of soil materials of human interest for excavation and (2) by the thematic and geographic limitations of the dataset. However, the attempt to consider the main human activities that require materials of soil origin and to map the locations where those materials are available on a continental scale provides new insight to this field of research.

These maps showing a continental overview can suggest opportunities for raw material extraction from soil which can serve current or future needs for a number of applications, such as construction and the health industry (soil organic matter). The maps show that northern and north-eastern Europe have large stocks for both applications, while most other regions of the continent do not have large reserves of soil organic matter but are generally well suited for extraction of construction materials, with substantial intraregional differences.

The comparison of maps showing the raw material availability of European soils (Figure 7) and productivity of various land uses (Figures 4, 5, and 6) also proves the need for separate assessment of the components of soil quality. In general terms the geographic locations of the most suitable zones for agriculture and for procuring soil-borne materials are different. On the other hand, there are more overlaps between prime areas of forest lands and zones with potential to provide raw materials from soil. While the first assumption is positive for resource use planning, the second highlights the dilemma that land use planners usually face: the utilization of one service of the land might negatively influence another utilization option. With the evaluation of other soil functions and ecosystem services, the level of complexity for decision-making increases.

CONCLUSIONS

A coherent view of soils requires an expansion of the conventional thinking about the role of soils and consideration of soils in a multidimensional perspective in the context of sustainable development including the utilization of soil-based ecosystem services. Conscious utilization of soil ecosystem services can only be achieved on the basis of a proper accounting.

In this study, an assessment of soil functions in relation to ecosystem services was carried out for a continental-scale overview of the EU. A coherent system of soil function-based ecosystem services was devised, taking into account the major soil functions as identified in the Soil Thematic Strategy.

In this attempt to characterize soil ecosystem services for the EU in a spatially explicit manner, new data on soil-based provisioning ecosystem services were produced, including productivity and raw material availability. Results show that up to an average of 60% of grassland productivity differences can be attributed to climatic factors between the most favorable (Atlantic) and least favorable (semi-arid Mediterranean) regions of the EU. While crop productivity shows a general trend toward increasing in a northward and westward direction, local soil quality in most regions—except in the Mediterranean—can compensate for climatic handicaps to a great extent. The attempts to address the main human activities requiring materials of soil origin and to map the locations where those materials are available on a continental scale provide new insight to this field of research.

The comparison of areas with potential for providing ecosystem services by individual soil functions highlights the complexity of decision-making for resource utilization but also the possibilities for optimization and more conscious management.

With the evolution of time-series soil data based on monitoring, the increase in spatial precision of soil information, and the introduction of dynamic components to the models, comparative assessment of soil functions can be a valuable tool for decision-making in the future across geographical scales and for different stakeholder's needs.

AUTHORS' CONTRIBUTIONS

GT, CG, AJ, SJ, LM, and TP participated in the concept development. GT, CG, and AJ designed the models. KB and EA contributed to the GIS analysis. GT and ÉI designed and implemented the model validation. GT, AJ, CG, and SJ drafted the manuscript. All authors read and approved the final manuscript.

AUTHOR'S INFORMATION

Gergely Tóth is a senior scientist at the Joint Research Centre, Italy, where his key research themes include investigations of soil quality, soil degradation processes, geographical distribution of soil properties, and soil nutrient dynamics. He is vice chair of the Soil Evaluation and Land Use Planning Commission of the International Union of Soil Sciences.

REFERENCES

1.   Aber JD, Ollinger SV, Driscoll CT (1997) Modelling nitrogen saturation in forest ecosystems in response to land use and atmospheric deposition. Ecol Model 101:61-78

2.   Bicheron P, Leroy M, Brockmann C, Krämer U, Miras B, Huc M, et al. (2006) GLOBCOVER: a 300m global land cover product for 2005 using ENVISAT/MERIS time series. Valencia.

3.   Blum WEH (2005) Functions of soil for society and the environment. Rev Env Sci Biotechnol 4(3):p75-p79

4.   Bouma E (2005) Development of comparable agro-climatic zones for the international exchange of data on the efficacy and crop safety of plant protection products. EPPO Bull 35:233-238

5.   Brunsdon C, Fotheringham AS, Charlton M (1996) Geographically weighted regression: a method for exploring spatial nonstationarity. Geogr Anal 28:281-289

6.   CEC (Commission of the European Communities) (2006) Communication from the Commission to the Council, the European Parliament, the European Economic and Social Committee and the Committee of the Regions—Thematic strategy for soil protection, Commission of the European Communities. Brussels: COM. p 231

7.   De Groot RS, Wilson MA, Boumans RMJ (2002) A typology for the classification, description and valuation of ecosystem functions, goods and services. Ecol Econ 41:393-408

8.   EC (2003) European soil database (distribution version v2.0). Ispra, Italy: European Commission Joint Research Centre.

9.   EEA (2006) Land accounts for Europe 1990–2000. Towards integrated land and ecosystem accounting. EEA Report No 11/2006. Copenhagen: EEA European Environmental Agency.

10.   Eurostat (2011) NUTS—Nomenclature of territorial units for statistics.   http://epp.eurostat.ec.europa.eu/portal/page/portal/nuts_nomenclature/introduction

11.   Eurostat (2013) Agricultural statistics - Crops products. http://epp.eurostat.ec.europa.eu/portal/page/portal/agriculture/introduction

12.   FAO (1976) A framework for land evaluation. FAO Soils Bulletin 32, Rome.

13.   FAO/UNESCO/ISRIC (1990) Revised legend of the soil map of the world. World soil resources report. Rome: FAO.

14.   Fisher B, Turner RK, Morling P (2009) Defining and classifying ecosystem services for decision making. Ecol Econ 68:643-653

15. Fotheringham AS, Brunsdon C, Charlton M (2002) Geographically weighted regression: the analysis of spatially varying relationships. Chichester: Wiley.

16. Hannam I, Boer B (2004) Drafting legislation for sustainable soils: a guide. Gland: IUCN.

17. Hartwich R, Baritz R, Fuchs M, Krug D, Thiele S (2005) Erläuterungen zur Bodenregionenkarte der Europäischen Union and ihrer Nachbarstaaten 1:5,000,000 (version 2.0). Hannover: Bundesanstalt für Geowissenschaften und Rohstoffe (BGR).

18. Haygarth PM, Ritz K (2009) The future of soils and land use in the UK: Soil systems for the provision of land-based ecosystem services. Land Use Pol 26(suppl1):S187-S197

19. Hellden U, Tottrup C (2008) Regional desertification: a global synthesis. Glob Planet Change 64:169-176

20. JRC-EEA (2005) CORINE land cover updating for the year 2000: image 2000 and CLC2000. Products and methods. Report EUR 21757 EN. Ispra: JRC.

21. Karlen DL, Andrews SS, Doran JW (2001) Soil quality: current concepts and applications. In: Sparks DL (ed) Advances in agronomy, 74, San Diego: Academic. pp 1-40

22. Köppen W (1936) Das geographische System der Klimate. In: Koppen W, Geiger R (eds) Handbuch der Klimatologie, Berlin: IC.

23. Larson WE, Pierce FJ (1991) Conservation and enhancement of soil quality. In: Evaluation for Sustainable Land Management in the Developing World, Vol. 2: Technical papers. Bangkok, Thailand: International Board for Research and Management. IBSRAM Proceedings No. 12 (2). pp 75-203

24. MEA (Millennium Ecosystem Assessment) (2003) Ecosystems and human well-being: a framework for assessment. A report of the conceptual framework working group of the Millennium Ecosystem Assessment. Washington DC: Island Press.

25. Panagos P, Van Liedekerke M, Filippi N, Montanarella L (2006) MEUSIS: towards a new multi-scale European soil information system. ECONGEO, 5th European Congress on Regional Geoscientific Cartography and Information Systems. Barcelona (Spain). pp 175-177

26. Parton WJ, Stewart JWB, Cole CV (1988) Dynamics of C, N, P and S in grassland soils: a model. Biogeochem 5:109-131

27. Peccol E, Movia A (2012) Evaluating land consumption and soil functions to inform spatial planning. 3rd International Congress on Degrowth for Ecological Sustainability and Social Equity. Venice (Italy). pp 19-23

28. Penman HL (1948) Natural evaporation from open water, bare soil and grass. Proc R Soc Series A 193:120-146

29. Rabus B, Eineder M, Roth A, Bamler R (2003) The shuttle radar topography mission—a new class of digital elevation models acquired by spaceborne radar. Photogramm Rem Sens 57:241-262

30. Raich JW, Rastetter EB, Melillo JM, Kicklighter DW, Steudler PA, Peterson BJ, et al. (1991) Potential net primary productivity in South America: application of a global model. Ecol Appl 1:399-429

31. Reid WV, Mooney HA, Cropper A, Capistrano D, Carpenter SR, Chopra K, et al. (2005) Millenium Ecosystem Assessment synthesis report. Washington DC: Island Press.

32. Robinson DA, Hockley N, Cooper DM, Emmett BA, Keith AM, Lebron I, Reynolds B, Tipping E, Tye AM, Watts CW, Whalley WR, Black HIJ, Warren GP, Robinson JS (2013) Natural capital and ecosystem services, developing an appropriate soils framework as a basis for valuation. Soil Biol Biochem 57:1023-1033

33. Running SW, Gower S (1991) FOREST-BGC, a general model of forest ecosystem processes for regional applications. II. Dynamic carbon allocation and nitrogen budgets. Tree Physiol 9:147-160

Chapter 4

Life Cycle Impact of Rare Earth Elements

P. Koltun[1] and A. Tharumarajah[2]

[1]CSIRO Process Science and Engineering, Gate 5, Normanby Road, Clayton, VIC 3168, Australia

[2]EnviSolutions, 20 Galahad Crescent, Glen Waverley, Melbourne, VIC 3150, Australia

ABSTRACT

The diverse properties of rare earth elements have seen broad and growing applications in clean energy technologies, hybrid vehicles, pollution control, optics, refrigeration, and so on. This study presents a "cradle-to-gate" life cycle assessment of the energy use, resource depletion, and global warming potential resulting from the production of rare earth elements (REEs) using the Bayan Obo rare earth operation in Inner Mongolia, China, as a representative system. The study aggregates data from the literature, LCI databases, and reasonable estimations. A novel economic value-based allocation method for the multiple coproducts of the process is proposed. It is found that four of the high priced REEs scandium, europium, terbium, and dysprosium have very high GWPs from production relative to the rest. A mass-based allocation is also provided for comparison. Impacts on immediate local environment from waste streams that can be toxic are not included in this study.

INTRODUCTION

Rare earth elements (REEs) or RE metals are technically defined as the 15 elements in the lanthanide (La) series, yttrium (Y), and scandium (Sc). Y and Sc are considered REEs since they mostly occur in the same ore deposits and have similar chemical properties. A precise classification often used in extraction is as follows [1].

- Light rare earth elements (LREEs): lanthanum (La), cerium (Ce), praseodymium (Pr), neodymium (Nd), and promethium (Pm).

- Medium rare earth elements (MREEs): samarium (Sm), europium (Eu), and gadolinium (Gd).

- Heavy rare earth elements (HREEs): terbium (Tb), dysprosium (Dy), holmium (Ho), erbium (Er), thulium (Tm), ytterbium (Yb), lutetium (Lu), scandium (Sc), and yttrium (Y).

There are about 200 known rare-earth containing mineral deposits, mostly as carbonatites spread around the world. Contrary to a lay persons' understanding, REs are not rare in natural occurrence (cerium is more abundant than tin and yttrium is more abundant than lead), though REEs have much less tendency to become concentrated in exploitable ore deposits [2], in particular HREEs. This being so, only a few mineral species, such as bastnasite, monazite, and RE- bearing clay, have been recovered for commercial production. Bastnasite deposits in China and the United States constitute the largest percentage of the world's rare-earth economic resources [3].

REEs possess diverse nuclear, metallurgical, chemical, catalytic, electrical, magnetic, and optical properties. This has led to ever-increasing applications, some mundane (lighter flints, glass polishing) to others that are highly specific. Examples of the latter include catalysts in petroleum refining industry, as alloying agents (used to enhance the oxidation resistance of alloys) in metallurgical processes, in glass, phosphors, optics, permanent magnets, and electronics. Emerging and potential applications include using rare earths to absorb ultraviolet light in automotive glass, corrosion protection, and metal coatings in corrosive and salty environments. Futuristic applications of REE are in high-temperature superconductivity, safe storage, and transport of hydrogen for a posthydrocarbon economy [2]. For applications of individual REEs, see [4].

Only small quantities of REEs are used in most applications, for instance, 2 to 3 grams in a 20-watt phosphor lamp. However, clean energy applications such as high capacity nickel-metal hydride batteries in hybrid cars (in a third-generation Toyota Prius) contain on average ~0.6 kg/battery of La [5] and RE magnets for electric motors as light weight alternative to iron magnets require around 1 kg of Nd [6, 7]. A utility scale wind turbine, another green power source, uses more than a ton of heavy-duty and lightweight magnets, 700 pounds of which is neodymium [8]. As the worldwide implementation of clean energy and other applications increase, many studies have identified the supply of particular REEs (dysprosium, neodymium, terbium, yttrium, and europium) to be at risk in the short and medium term [9, 10].

China currently accounts for 95% of global production of rare earths (about 90% of REs used in the US) and domestic industry consumes about 60% [18]. Bayan Obo mine (used as the reference site in this study) situated in Inner Mongolia is said to supply half of the total production [19]. However, this situation may change due to China's introduction of progressively reducing export quotas since 2004/2005 [5] for reasons of environmental concerns and conservation of resources [20]. This potential risk to supply may be somewhat absorbed by recycling of in-use stocks of REEs in products (an estimated 440,000 metric tons in 2007, [21]). In fact, this source can also help to reduce the impact on the environment from the production of virgin REEs.

Many environmental issues surround the production and use of REEs. REEs having similar chemical structures are difficult to separate [22] compounded by extremely low yield (net recovery from the ore at Bayan Obo is around 0.6% [23, 24]). Thus, apart from the electricity, acids, water, and resources expended in production, there can be huge amounts of waste that can be toxic with potential damage to the ecosystem.

One of the most important environmental issues for producers of rare earths is the problem posed by the radioactivity of thorium-containing monazite and xenotime ores. Australian monazite exports declined precipitously between 1989 and 1992 in part because of the radioactivity concern and since 1992, Rhodia, at that time the world's largest rare earth company, has used Chinese instead of Australian rare earth concentrates for production of individual rare earth oxides [25].

Other concerns of RE pollution relate contamination of the local environment. These include potential bioaccumulation in the food chain from waste or dust contamination of water ways and soil from mines when used in downstream industrial processes and in agricultural fertilisers [27, 28].

This study, with the exception of impact on the local environment, investigates the global warming impact and intensity of use of energy and resources in the production of RE oxides (REOs) and REEs. Each of the distinct stages from mining, separation of REOs, and reduction to REEs is modelled and the input of resources (energy, water, chemicals) and wastes is examined. GHG impacts and resource depletion potential are computed from the derived data.

It is the objective of this study to disseminate knowledge of the GHG and other impacts of producing REEs, in particular at this time when their application is growing and such knowledge is not readily available. Further, it is believed that such data will be valuable and provide a basis for constructing and comparing GHG impact profiles of rare earth based products for comparison purposes.

EXTRACTION AND PRODUCTION OF RARE EARTHS

Sources and Distribution of Rare Earth Minerals

Only a few mineral species, such as bastnasite (a rare earth fluorocarbonate (Ce, La)(CO_3)F), monazite (a rare earth phosphate (Ce, La, Y Th) PO_4)), xenotime (YPO_4), and RE-bearing clay have been recovered for commercial production. Of these, bastnasites constitute the largest percentage (in US and China). Monazite deposits in Australia, Brazil, China, India, Malaysia, South Africa, Sri Lanka, Thailand, and the United States constitute the second largest segment. World distribution is shown in Figure 1. Undiscovered resources are thought to be very large relative to expected demand [3].

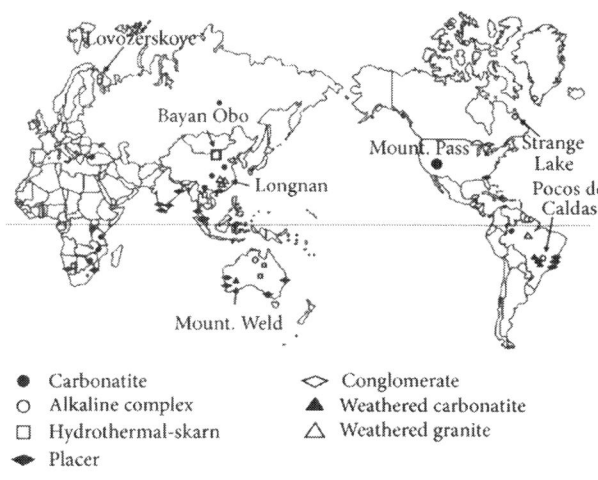

Figure 1: World distribution of rare earths [26].

Around 95% of the world's rare earth metals are produced in China, and as mentioned before Bayan Obo mine supplies about half of the total production. This mine is estimated to contain at least 1.5 billion metric tons of iron (average grade 33% of iron oxide), 89 million tons of RE oxides (REOs) (average grade 6%), and 1 million tons of niobium (average grade 0.13%) [23]. The principal RE minerals in Bayan Obo are bastnasite and monazite accompanied by RE-Nb minerals; further details of the geological distribution are given in [23]. This study uses Bayan Obo operations as a representative system for rare earth extraction and production. Rare earth production at Mount Pass in USA ceased operation in 2002 due to environmental concerns (production has recommenced since 2008). The projected production in 2012 is expected to be between 8,000 and 10,000 metric tons [29]. Mining at the Mount Weld deposit in Western Australia began in 2007, and an estimated 98,000 cubic meters of ore has been stockpiled awaiting the completion of a concentration plant at the mine site. The concentrates will be exported to an advanced materials plant being built in Malaysia [30].

REEs usually have very small differences in solubility and complex formation and hence their separation can be difficult [22]. Thus, separation processes are often chemically intensive using ion exchange methods, solvent extraction, or fractional crystallization. The average recovery rate from RE containing ores is extremely low at 10% [24]. This is compounded by the low grade of REO in mined minerals at Bayan Obo (around 6% [23]). Thus, the net recovery is about 0.6% and results in high resource inputs (electricity, acids, water and resources expended in production) as well as huge amounts of waste that can be toxic with potential damage to the ecosystem.

A representative system of extraction and production in Bayan Obo is shown in Figure 2. This representation is used as the system boundary for this LCA study and is explained briefly.

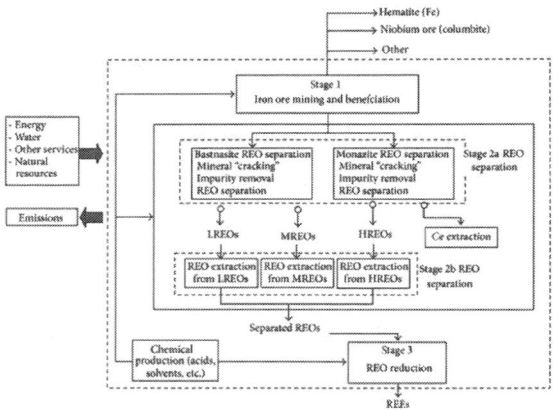

Figure 2: System boundary: mining and production of RE metals in China.

Stage 1: Mining and Beneficiation

Bayan Obo mineral Fe-REO-Nb deposit in China is the source of both iron-ore and RE containing minerals bastnasite and monazite. In separating the concentrates of bastnasite and monazite, a combination of separation processes (after grinding) are used including wet-magnetic separation and froth floatation. Since mining and beneficiation are established technologies, data is mostly based on our previous studies in mining and beneficiation [31, 32].

Stage 2: Purification and Separation of REO

The concentrates of bastnasite and monazite (approximately 85% concentration by weight within the slurry [24]) from mining and beneficiation separately undergo further processing (shown as Stage 2a in Figure 2). This stage purifies and separates the light, medium, and heavy oxide groups using mineral cracking, acid leaching, impurity removal, and precipitation. While Stage 2a processes are similar for both concentrates, due to the substantial amount of radioactive elements in monazite (together thorium and uranium account about 6.5% of monazite) and phosphorus (phosphorus oxide accounts about 28% of monazite), the monazite processing is more energy and chemical intensive than bastnasite.

The extraction of RE oxide groups from both bastnasite and monazite produces cerium and separate streams of light, medium, and heavy RE oxides. The separated streams are further processed to produce the individual REOs. To extract the individual REOs, the oxide groups are subject to multistage extraction process using extractants in acidic medium. LREO separation involves 22 stages: 8 for extraction, 8 for scrubbing, and 6 for stripping [33]. MREOs have 29 extraction stages and 44 scrubbing stages [34]. Currently heavy RE oxides are separated by using ion exchange process with reagent impregnated resins [35].

Stage 3: Reduction of REEs

The industrial reduction to REEs from light REOs (La, Nd) including mischmetal (which are a mixture of light lanthanide metals) uses fused-salt electrolysis (fused salt process is used when the reactivity of the metal does not allow electrowinning from aqueous solutions).

For MREEs and HREEs, metallothermic process using electrolysis from an aqueous solution or simple carbon based pyrometallurgy is employed due to the high electropositive nature of RE elements in these groups [36].

Processes for disposing toxic and other waste from the mining and processing are not considered in this study.

LCA STUDY METHODOLOGY AND ASSUMPTIONS

The principal goal of this study is to investigate the cradle-to-gate energy use (assessed as electricity use, heating fuel use, and total nonrenewable energy use), water use, and global warming impact of REEs produced in Bayan Obo, China. It is envisaged that the impact data of REEs created in this study could provide a basis for constructing and comparing impact profiles of rare earth based products.

On defining the scope of this study, the system boundary definition for ascertaining the impact is shown in Figure 2 and is explained in Section 2.2. Other topics of scope such as allocation, data sources, and uncertainty treatment are discussed below. Results of impact assessment are given in Section 4.

One aspect of scope that is given careful attention is the allocation of impacts. Normally, extraction of REEs is a multifunctional product system, meaning a number of outputs as saleable products are produced at the different stages of production. For instance, in Bayan Obo mining operation, hematite, niobium, and RE containing ores are produced as saleable coproducts of mining. The separated RE containing ore is then processed into saleable REOs or can be further reduced to produce individual REEs. Thus, properly allocating from "cradle to gate" (i.e., from mining to when REOs or reduced REEs are sold) is important.

Allocation among the coproducts uses both mass fractions (i.e., physical relationship) as well as an economic model. In fact, while mass-based allocation is sufficient as recommended by ISO 14044 [37], an additional economic model that combines mass and price basis (market value) is proposed here. This model reflects the underlying causality between economic reasons for producing coproducts (i.e., REEs from iron ore extraction) and their environmental impact as noted by [38]. It has also been suggested that economic allocations may be suited for coproducts that vary in prices [39] as is the case for REOs.

At Stage 1 iron ore mining (see Figure 2), three economically significant commodities (coproducts) are mined: hematite (FeO), columbite (niobium ore), and REO bearing ore (i.e., bastnasite and monazite minerals). Mass-based allocation uses the mass fractions of each product (see Table 1).

Table 1: Bayan Obo iron ore composition and relative economic value of coproducts

Commodity	Fraction in Bayan Obo deposit, %	P_c—Price per tone*, US$	C_c—Normalised fraction, %	X_c—Share in economic value, %
Iron ore (hematite)	33	120.35	84.33	2.87
RE oxides (RE_xO_y)	6	215400.00	15.33	93.31
Niobium ore (columbite)	0.13	41000.00	0.33	3.82
Fluorite (CaF_2)	2	—	—	0
Thorium oxide (ThO_2)	0.16	—	—	0
Phosporus oxide (PO_4)	0.86	—	—	0
Rest	57.85	—	—	0

*Prices are shown for the year 2009-2010. Prices estimated for iron ore [11], REO [12], and niobium ore [13].

Economic allocation among these ores is performed on the basis of share of contribution of each product according to (1), where C_c, P_c, and X_c are, respectively, the mass fraction, price, and computed environmental share of commodity c. Mass allocation uses C_c as the basis. Share of allocation for each of the mined coproducts is shown in Table 1. The mass fractions (C_c) are adjusted (normalized) across the economically valuable ores (iron ore, REO, and niobium):

$$X_c = \frac{P_c * C_c}{\sum_c (P_c * C_c)}.$$

(1)

In Stages 2a and 2b, the plant produces a mixture of REOs as coproducts. This would require calculation of impact that can be assigned to each of the individual REOs. A combined mass and economic allocation model similar to Stage 1 is developed and applied for this as described by the following equations:

$$C_i = c_b b_i + c_m m_i,$$

$$(2)$$

$$X_i = \frac{P_i * C_i}{\sum_i (P_i * C_i)}.$$

$$(3)$$

In (2), the average mass composition (C_i, used for mass-based allocation) of REO_i is determined using itsmass composition in bastnasite (b_i) and monazite (m_i) weighted according to the ratio of bastnasite (c_b) and monazite (c_m) processed. The proportion of bastnasite and monazite in Bayan Obo mineral deposit is taken as $3:1$, respectively, with each containing 60% of RE oxides [40].

Equation (3) computes the share of the environmental burden (X_i) using price per kg (P_i) of REO_i as the weighting factor normalised over all REOs. This way, both mass and economic proportions are combined to derive the final allocation given to each REO.

The values of pertinent parameters in (2) and (3) used in calculating the share of the environmental burden assigned to the REOs are given in Table 2. References used appear as notes in the table. The following REOs are not considered.

Table 2: Share of environmental impact allocated for REOs

REE	REO	i	, %	, %	, %	, US$/kg	, %
Lanthanum	La_2O_3	La	23.00	23.00	23.00	13.00	7.75
Cerium	CeO_2	Ce	50.00	46.00	49.00	12.00	15.24
Praseodymium (+3)	Pr_6O_{11}	Pr	6.20	5.00	5.90	95.00	14.53
Neodymium	Nd_2O_3	Nd	18.50	19.00	18.63	77.00	37.18
Promethium	Pm_2O_3	Pm	—	—	0.00	0.00	0.00
Samarium	Sm_2O_3	Sm	0.80	3.00	1.35	23.50	0.82
Europium	Eu_2O_3	Eu	0.20	0.10	0.18	2150.00	9.75
Gadolinium	Gd_2O_3	Gd	0.70	1.70	0.95	75.00	1.85
Terbium	Tb_4O_7	Tb	0.10	0.16	0.12	1750.00	5.22
Dysprosium	Dy_2O_3	Dy	0.10	0.50	0.20	975.00	5.05

Holmium	Ho_2O_3	Ho	0.00	0.00	0.00	—	0.00
Erbium	Er_2O_3	Er	0.00	0.13	0.03	77.00	0.06
Thulium	Tm_2O_3	Tm	0.00	0.00	0.00	—	0.00
Ytterbium	Yb_2O_3	Yb	0.00	0.06	0.02	69.12	0.03
Lutetium	Lu_2O_3	Lu	—	—	0.00	—	0.00
Scandium	Sc_2O_3	Sc	—	0.04	0.01	7200.00	1.87
Yttrium	Y_2O_3	Y	0.00	2.00	0.50	50.00	0.65

References used: b_i—[14]; m_i—[1], Sc_2O_3—[15]; P_i—[16], Sm_2O_3, Yb_2O_3—[17].

Tm and Ho are least abundant and are found with other rare earths such as gadolinite and other minerals containing rare earths, and Lu is the rarest of the rare earths [41]. Due to data availability, these are not considered here.

Pm is the only radioactive rare-earth metal of transition Group IIIb of the periodic table and not detected in nature [42]. It has a low period of half decay and its main route of production is artificial synthesis.

In Stage 3 REO-REE reduction, the mass fraction of individual REE present in the corresponding RE oxide is used for allocation.

The manner in which the above methods are used to calculate the impact allocated to REO or REE products, covering mining to production life cycle stages, is described below.

First, for the production of REO products, a reference flow for producing 1 kg mixture of REO products is considered at Stage 2. To obtain this, 166.7 kg of iron ore has to be mined (using average REE grade of 6% in the ore and REO recovery rate of 10%). Impact of mining this (Stage 1) is allocated among Fe, Nb, and RE containing concentrates (i.e., bastnasite and monazite) as described.

For Stage 2a (separation of REO groups), share of allocation derived for each REO is used (Table 2). In the case of Stage 2b (separation of REOs from REO groups), the same model as used for Stage 2a is used, though it is normalized for the REOs for the oxides occurring within each group. Impact from mining to REO production is the sum of the Stages 1–2 for the fraction of REOs in the reference flow. Finally, the impact per kg of individual REO is calculated.

In deriving the allocation for each REE, the impact of REO-REE reduction stage is first assigned on a mass basis, that is, the mass fraction of individual REE present in the corresponding RE oxide. "Cradle to gate" impact per kilogram of individual REE is determined by adding the impact of corresponding REO and this stage.

Inventory data for the processes is obtained from publicly available information sources (cited appropriately under each section) and by estimation. These include review of technological processes for RE production and analogous processes, environmental data pertaining to materials and chemicals mainly from EcoInvent life cycle inventory (LCI) database [43], and other LCI databases combined with modeling and estimation. Table 3provides the basic data for energy and water use at each stage from mining to REO extraction. The diverse sources of data require uncertainty analysis, such as Pedigree Matrix [44] combined with Monte Carlo simulation available in SimaPro software [45]. However, such analysis will be performed as a future extension of this study.

Table 3: Life cycle inventory and impact of production stages for the production of 1 kg of RE oxide

Stage	Process	Energy consumption, MJ		Water consumption, kL	Environmental impact	
		Electricity	Heat energy		GHG emissions, kg CO_2 eq.	Resource depletion, MJ surplus
Stage 1	Mining and beneficiation	4.64	10.11	3.24	3.95	5.46
Stage 2a	REO extraction from bastnasite	5.60	90.00	19.09	10.5	13.9
	REO extraction from monazite	55.60	11.90	18.15	18.3	19.6
Stage 2b	Light REO separation	12.60	0	0.58	3.3	2.80
	Medium REO separation	5.20	0	0.16	1.3	1.29
	Heavy REO separation	15.50	0	8.52	5.5	5.57

The major assumptions of this study include the following.

• Distance between mining pit and beneficiation plant and beneficiation and REO separation plant are assumed as 15 km and 30 km, respectively. REO reduction and subsequent reduction to

REE occur at the same place, that is, Bayan Obo Township.

- In separating REOs from monazite concentrates radioactive wastes can occur. Treatment of such waste can pose additional environmental effects including toxicity and soil and ground water contamination. The treatment processes of such waste are considered out of scope.

- Average grid-mix of electricity in China is 75% from coal power stations and the rest is from hydro [31].

- Purity range of REOs produced (to derive the life cycle inventory data) is 98.0%–99.9% associated with current production technology in China.

- A steady-state process with a constant rate of materials and energy flows is assumed.

- Chemical and other materials used in the processes are assumed to be imported from Europe and hence, the energy consumption for their production is mostly based on data taken from the European LCI databases ([43, 45]).

ENVIRONMENTAL IMPACT

The life cycle inventory (LCI) of energy (both electricity and heat) and water consumption and corresponding global warming impact for mining and separating REOs (i.e. stages 1 and 2 in Figure 2) to produce 1 kg of mixed REOs are shown in Table 3. Assignment of impacts among products uses the allocation methods described already. The widely used Eco-indicator 99 impact assessment methodology [46] is used in selecting and computing the appropriate environmental damage indicators. SimaPro software [45] is used in deriving the indicator scores.

Two indicators selected for this study are GHG for global warming potential and resource depletion. Resource depletion is defined in the Eco-indicator 99 methodology report [46]. It is measured in terms of surplus energy (MJ) required to obtain the same quality of resource as they become depleted. The calculation of this indicator uses geostatistical models to analyse the concentration of a mining resource at a reference year followed by assumptions on future depletion rates. The somewhat arbitrary and limiting nature of assumptions and calculation makes this indicator useful only as a comparative measure (the methodology report in [46] provides more information). Nevertheless, in the absence of more objective indicators resource depletion is used here.

Examining Table 3, one can see that energy consumption and GHG emissions for the REO extraction (Stage 2a) are much more for monazite than bastnasite, due to the required additional separation of uranium, thorium, and phosphorous for the former. In Stage 2b, the impact is the highest for separation of HREOs due to ion exchange process used. While solvent extraction process is used for both LREOs and MREOs with the latter being more difficult, impact from LREO extraction is higher due to the higher mass contribution (1 kg mixed REOs contains about 96% LREOs). Out of all the stages, Stage 2a (REO extraction, see Table 3) has the most resource deletion of around 33.5 MJ surplus

Impact for each of the separated REOs from Stages 1 and 2 is computed using both mass- and price-based allocation methods described in the previous section. Price-based allocation of impact per kilogram of each REO is shown in Table 4.

Table 4: Life cycle inventory and impact for the production of 1 kg RE oxides using price-based allocation

RE oxide	Classification	Energy consumption		Water consumption, kL	Environmental impact	
		Electricity, MJ	Heat, MJ		GHG emissions, kg CO_2eq.	Resources depletion, MJ surplus
La_2O_3	Light	26.13	34.39	12.82	11.16	12.52
CeO_2	Light	24.12	31.74	11.83	10.30	11.56
Pr_6O_{11}	Light	190.97	251.28	93.69	81.53	91.52
Nd_2O_3	Light	154.78	203.67	75.94	66.09	74.18
Pm_2O_3	Medium	0.00	0.00	0.00	0.00	0.00
Sm_2O_3	Medium	37.95	62.16	22.74	17.73	20.59
Eu_2O_3	Medium	3472.18	5686.98	2080.32	1622.04	1883.96
Gd_2O_3	Heavy	121.12	198.38	72.57	56.58	65.72
Tb_4O_7	Heavy	2826.83	4628.93	1718.02	1325.42	1538.92
Dy_2O_3	Heavy	1574.95	2578.98	957.18	738.45	857.40
Ho_2O_3	Heavy	0.00	0.00	0.00	0.00	0.00
Er_2O_3	Heavy	124.38	203.67	75.59	58.32	67.71
Tm_2O_3	Heavy	0.00	0.00	0.00	0.00	0.00
Yb_2O_3	Heavy	111.65	182.83	67.86	52.35	60.78
Lu_2O_3	Heavy	0.00	0.00	0.00	0.00	0.00
Sc_2O_3	Heavy	11630.4	19044.8	7068.42	5453.15	6331.55

Y_2O_3	Heavy	80.77	132.26	49.09	37.87	43.97

Comparison of environmental impacts among the REOs shows that LREOs (La-Nd) have lower impact than the rest of the REOs with La and Ce being minimal within the LREO group. The impact for other REOs varies substantially and cannot be attributed to their classification as MREOs and HREOs.

The impacts assigned to REOs are influenced by the concentration and price; the lower the percentage concentration and the higher the price, the higher the share of impact per kg. Among the REOs, scandium oxide (Sc_2O_3) has the highest GHG impact at ~6,332 kg of CO_2 eq./kg due to its very high price (USD 7,200 per kg) and very low occurrence (0.01%). The next highest impact is for europium oxide (Eu_2O_3) at ~1,884 kg of CO_2 eq./kg. Its concentration is comparatively higher and price is lower than Sc_2O_3. Other notable high-impact REOs are terbium oxide (Tb_4O_7, ~1,539 kg of CO_2 eq./kg) and dysprosium oxide (Dy_2O_3, ~857 kg of CO_2 eq/kg).

When mass-based allocation is used, a separated REO carries the same impact assigned to an RE group. This is due to the allocation being normalized by group and derived on a per kg basis. Table 5 shows the mass-based impact by RE group. Here, the values in the table are affected by Stage 2b, which is particular to the groups. Hence, MREOs, being lower than others, have lower inventory and impact values.

Table 5: Life cycle inventory and impact for the production of 1 kg RE oxides using mass-based allocation

Classification	Energy consumption		Water consumption, kL	Environmental impact	
	Electricity, MJ	Heat, MJ		GHG emissions, kg CO_2 eq.	Resources depletion, MJ surplus
Light REOs	74.08	102.51	38.06	32.29	36.50
Medium REOs	66.68	102.51	37.64	30.29	34.99
Heavy REOs	76.98	102.51	37.64	34.49	39.27

A comparative GHG profile of REOs with allocations based on price and mass is shown in

Figure 3. As noted, GHG allocation using mass-based allocation is the same for REOs in the same group.

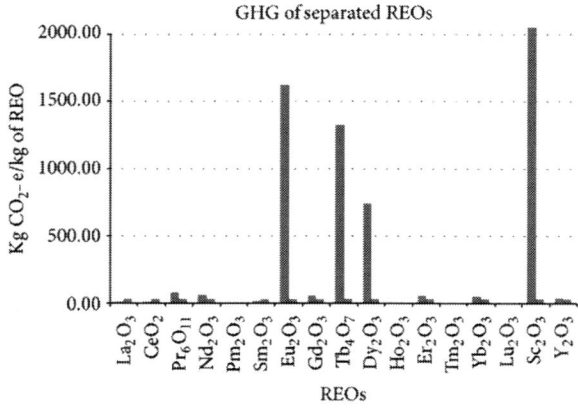

Figure 3: Comparative GHG of rare earth oxides for mass- and price-based allocations.

Separated REOs are further processed in a reduction stage to extract the individual REEs. This study assumes that the reduction plants are located in Bayan Obo and no or minimal transport takes place. An average impact is calculated for electrolytic process (for LREEs) and metallothermic process (HREEs and MREEs) based on [47] and [48], respectively. Thus, the impact calculated for this stage alone is affected by the process yield of 80% for LREEs and 95% for other REEs; the impacts for reduction are, respectively, 5.64 and 5.04 kg of CO_2eq./kg of REE.

DISCUSSION

GHG emissions from "cradle-to-gate" life cycle stages of production (i.e., mining, REO separation, and REE reduction) based on mass- and price-based allocation for extracting 1kg of individual REE are displayed in Figure 4. Mass-based approach allocates more emissions to more abundant elements, though when computed per kilogram of REE, the mass proportion becomes insensitive. Thus, the factor that influences the comparative allocation is the energy and materials expended (differs between RE groups) and the extractable mass of REE from its oxide. This is apparent in Figure 4(a) where Sc has the highest impact followed by Y. The extractable mass of REE from the 25 oxide for these two is, respectively, 65% and 78% compared to others that are in the range of 81%–85%.

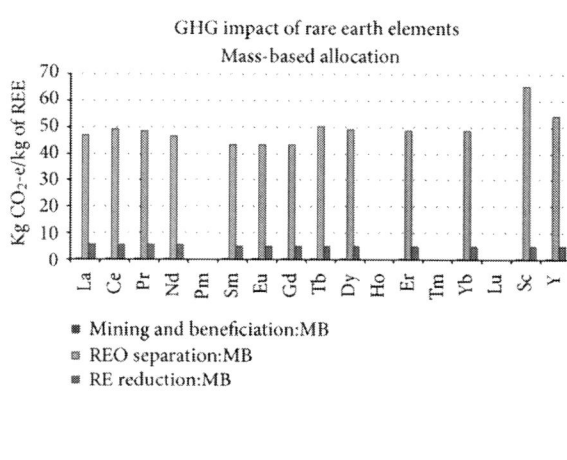

(a)

(b)

Figure 4: Cradle-to-gate GHG emissions of rare earth elements.

Unlike mass-based allocation, price-based allocation model proposed here uses both mass concentrations in mineral and price. The comparative impacts allocated are hence influenced by: (1) energy and materials consumed at each stage; (2) extractable mass of REE in the minerals, that is concentrations in bastnasite and monazite and mass proportion in the respective oxides; and (3) price used in calculating the share of environmental burden to be allocated.

In this model, scandium, europium, terbium, and dysprosium have relatively very high impact (Figure 4(b)). The corresponding REOs of these are high too for reasons cited in the last section. In reducing them to REEs, a further increase in the case of scandium occurs due to its lower extractable mass from its oxide of ~65% compared to ~86% for europium, ~85% for terbium, and ~87% for dysprosium. Thus, to produce 1 kg of Sc would require 30% more Sc_2O_3 than other oxides. Additionally, Sc_2O_3, Tb_4O_7, and Dy_2O_3 are HREOs, the separation of which requires more energy. Economic allocation is influenced by price fluctuations. While the prices of REOs have been volatile dropping as much as 40%–60% in 2011-12 [16], nevertheless, the comparative prices of the above four high impact REOs have been consistently higher than other REOs. Thus, they remain the high impact rare earths relative to others. In any case, the impact of individual REEs can vary with price fluctuations.

Comparison of the two allocation methods shows that mass-based allocation is not sensitive to the scarcity of the rare earth in the mineral (per kilogram basis). The suggested allocation based on combination mass, concentration, and price shows such sensitivity; for example, allocated GHG emissions for low concentration REEs and those that are highly priced tend to be high.

Further, Bayan Obo is unique as an iron ore mine that also produces rare earths, and though similar, Mountain Pass does not produce iron ore or niobium. This means that in this case only part of the total energy and environmental impacts associated with mining stage is assigned to rare earths. While the impact of this stage may be smaller than other stages, this peculiarity of the model has to be recognised.

CONCLUSIONS

This study has in some detail investigated the cradle-to-gate GHG impact of a rather complex route of producing REEs starting from extraction of rare earth minerals, separation of oxides, and final reduction of REEs.

Both mass- and price-based allocation models have been employed in estimating the impact. The former is only sensitive to the extractable mass concentrations, whereas price based allocation model proposed in this study is additionally sensitive to price. Thus, where the price of coproducts of product systems varies widely as in the case of rare earths, it tends to amplify the impact of highly priced rare earths that have lower concentrations. This information can be useful in focussing efforts to improve process efficiency and recycling to increase supply.

Recycling is an attractive pathway given the increasing prices of rare earths (while 2011-12 prices were down, they are generally higher than 2010 prices) combined with the recent clamp down on exports by China. The route to recycling can be closed-loop, meaning the recovery of the original RE alloys with minimum loss of property for similar applications. Such direct recycling, however, has its challenges in collecting, sorting, separating components, and finding suitable processes. An open-loop recycling, where REEs can be recovered from alloys for use in other applications, is also viable. Investigation of GHG impacts of the various routes to recycling is planned as the extension of this study.

An additional concern is the impact on the environment from processing waste. In particular, the large amount of tailings is produced in beneficiation and extraction of bastnasite and monazite concentrates. Tailings of both of these, monazite in particular, contain naturally occurring radionuclides and the release of this to the environment by air, wastewater, and rain leaching can have longer-term health effects to humans and ecosystems of the local environment. An assessment of these and other impacts from waste processing and disposal is seen as supplementing this study.

REFERENCES

1. Australian Industry Commission, New and Advanced Materials, Australian Government Publishing Service, Melbourne, Australia, 1995.

2. U.S. Geological Survey, "Rare Earth Elements-Critical Resources for High Technology," Fact Sheet 087-02, 2002, http://pubs.usgs.gov/fs/2002/fs087-02/.

3. U.S. Geological Survey, "Mineral Commodity Summaries—Rare Earths," 2008,http://minerals.usgs.gov/minerals/pubs/commodity/rare_earths/mcs-2008-raree.pdf.

4. HEFA Rare Earth Canada Co. Ltd, Rare Earth products by element, http://www.baotou-rareearth.com/.

5. U.S. Department of Energy, "Critical Materials Strategy," December 2011, http://energy.gov/pi/office-policy-and-international-affairs/downloads/2011-critical-materials-strategy.

6. S. Gorman, "As hybrid cars gobble rare metals, shortage looms," Reuters, 30th August 2009,http://www.reuters.com/article/GCA-GreenBusiness/idUSTRE57U02L20090831.

7. S. Gorman, "California mine digs in for, "green" gold rush," Reuters, 30th August 2009,http://www.reuters.com/article/idUSTRE57U02I20090831?sp=true.

8. A. Pasternack, "China Tightens Grasp on Rare Earth Metals Vital for Green Technologies," Treehugger (Business & Politics), August 2009, http://www.earth-stream.com/outpage.php?s=18&id=201861.

9. D. Bauer, D. Diamond, J. Li, D. Sandalow, P. Telleen, and B. Wanneret, "Critical Materials Strategy, US Department of Energy," 2010,http://energy.gov/sites/prod/files/edg/news/documents/criticalmaterialsstrategy.pdf.

10. T. E. Graedel, "On the future availability of the energy metals," Annual Review of Materials Research, vol. 41, pp. 323–335, 2011.

11. "Iron Ore Monthly Price," November 2012, http://www.indexmundi.com/commodities/?commodity=iron-ore&months=12.

12. D. Kingsnorth, "Rare earths: is supply critical in 2013?" in Proceedings of the Critical Minerals Conference Perth (AusIMM '13), Curtin Graduate School of Business; Curtin University & Industrial Minerals Company of Australia Pty Ltd, Western, Australia, June 2013.

13. U.S. Geological Survey, "Niobium (Columbium) and Tantalum Statistics and Information," Mineral Commodity Summaries— Niobium (Columbium) 2012,http://minerals.usgs.gov/minerals/pubs/commodity/niobium/mcs-2012-niobi.pdf.

14. U.S. Geological Survey, Rare Earths Statistics and Information 2010, Minerals Yearbook 2010—Rare Earth Contents of Major and Potential Source Minerals, 2010,http://minerals.usgs.gov/minerals/pubs/commodity/rare_earths/myb1-2010 raree.xls.

15. C. Yan, J. Jia, C. Liao, S. Wu, and G. Xu, "Rare earth separation in China," Tsinghua Science and Technology, vol. 11, no. 2, pp. 241–247, 2006.

16. MineralPrices, December 2012, http://www.mineralprices.com/.

17. AsianMetal, "Rare Earth Prices," December 2012, http://www.asianmetal.com/price/initPriceListEn.am?priceFlag=8&productInfoIMG=PrNd+Mischmetal%E3%80%8299%EF%BC%85min+Nd+75%EF%BC%85+China%E3%80%82RMB%2FMT#.

18. The China Post, "China will limit its exports of rare earth metals, official says," 3rd August 2009,http://www.chinapost.com.tw/business/asia/b-china/2009/09/03/223211/China-will.htm.

19. W. M. Morrison and R. Tang, "China's Rare Earth Industry and Export Regime: Economic and Trade Implications for the United States," Congressional Research Service 7-5700 R42510, April 2012,http://www.fas.org/sgp/crs/row/R42510.pdf.

20. ChinaDaily, "China announces rare earth export quotas," July 2011,http://www.chinadaily.com.cn/china/2011-07/14/content_12906607.htm.

21. X. Du and T. E. Graedel, "Global in-use stocks of the rare earth elements: a first estimate,"Environmental Science and Technology, vol. 45, no. 9, pp. 4096–4101, 2011.

22. A. M. Helmenstine, "Rare Earth Properties," 2009,http://chemistry.about.com/od/elementgroups/a/rareearths.htm.

23. L. J. Drew, M. Qingrun, and S. Weijun, "The Bayan Obo iron-rare-earth-niobium deposits, Inner Mongolia, China," LITHOS, vol. 26, no. 1-2, pp. 43–65, 1990.

24. C. W. Sinton, "Study of the Rare Earth Resources and Markets for the Mt," Weld Complex for Lynas Corporation. BCC Research, Washington, DC, USA, 2008,http://www.lynascorp.com/content/upload/files/press_releases/BCC_FINAL_REPORT.pdf.

25. R. Will, A. Eric, and N. Takei, "Chemical Economics Handbook—Rare Earth Minerals and Products," SRI Consulting, Menlo Park, Calif, USA, 2010,http://www.sriconsulting.com/CEH/Public/Reports/765.5000/.

26. Y. Kanazawa and M. Kamitani, "Rare earth minerals and resources in the world," Journal of Alloys and Compounds, vol. 408-412, pp. 1339–1343, 2006.

27. G. X. Xing and H. M. Chen, "Environmental impacts of metal and other inorganics on soil and groundwater in China," in Soils and Groundwater Pollution and Remediation: Asia, Africa, and Oceania, P. M. Huang and I. K. Iskandar, Eds., pp. 167–200, CRC Press, 1999.

28. S. Zhang and X.-Q. Shan, "Speciation of rare earth elements in soil and accumulation by wheat with rare earth fertilizer application," Environmental Pollution, vol. 112, no. 3, pp. 395–405, 2001.

29. MolyCorp, "Mountain Pass Production," 2011, http://www.molycorp.jp/about-us/current-future-production/.

30. Geoscience Australia, "Rare Earth Production and Exports," December 2012,http://www.ga.gov.au/minerals/mineral-resources/rare-earth-elements.html.

31. S. Ramakrishnan and P. Koltun, "Global warming impact of the magnesium produced in China using the Pidgeon process," Resources, Conservation and Recycling, vol. 42, no. 1, pp. 49–64, 2004.

32. A. Tharumarajah and P. Koltun, "Is there an environmental advantage of using magnesium components for light-weighting cars?" Journal of Cleaner Production, vol. 15, no. 11-12, pp. 1007–1013, 2007. · ·

33. C. A. Morais and V. S. T. Ciminelli, "Process development for the recovery of high-grade lanthanum by solvent extraction," Hydrometallurgy, vol. 73, no. 3-4, pp. 237–244, 2004. · ·

34. L. Wenli, G. D. ‹. Ascenzo, R. Curini et al., "Simulation of the development automatization control system for rare earth extraction process: combination of ESRECE simulation software and EDXRF analysis technique," Analytica Chimica Acta, vol. 417, no. 1, pp. 111–118, 2000. · ·

35. C. Liao, C. Yan, and J. Jia, "A Novel Solvent Extraction System for Tm, Yb and Lu," China Patent CN, 98100226.9, 2008, http://www.sciencedirect.com/science/article/pii/S1007021406701833.

36. P. F. Duby, Kirk-Othmer Encyclopedia of Chemical Technology, John Wiley & Sons, 2005. ·

37. International Standards Organisation, "Environmental management—Life cycle assessment—Requirements and guidelines," ISO 14044, 2006.

38. K. Kodera, Analysis of allocation methods of bioethanol LCA, Internship at CML [M.S. thesis], Faculty of Earth and Life Science, Leiden University, Leiden, The Netherlands, 2007.

39. F. Ardente and M. Cellura, "Economic allocation in life cycle assessment: the state of the art and discussion of examples," Journal of Industrial Ecology, vol. 16, no. 3, pp. 387–398, 2012. · ·

40. J. Ren, S. Song, A. Lopez-Valdivieso, and S. Lu, "Selective flotation of bastnaesite from monazite in rare earth concentrates using potassium alum as depressant," International Journal of Mineral Processing, vol. 59, no. 3, pp. 237–245, 2000. · ·

41. Development Corp., Rare Earth Elements, http://www.candldevelopment.com/rare_earth_detail.htm.

42. C. R. Hammond, The Elements, Handbook of Chemistry and Physics, CRC press, 81st edition, 2000.

43. EcoInvent Centre, http://www.ecoinvent.org/database/.

44. B. P. Weidema and M. S. Wesnæs, "Data quality management for life cycle inventories—an example of using data quality indicators," Journal of Cleaner Production, vol. 4, no. 3-4, pp. 167–174, 1996.

45. Pre Consultants, 2011, http://www.pre.nl/.

46. Pre Consultants, "A Damage oriented method for Life Cycle Assessment, Methodology Report," 2nd edition, 2000, http://

teclim.ufba.br/jsf/indicadores/holan%20ecoindicator%2099. pdf.

47.   E. S. Shedd, J. D. Marchant, and T. A. Henrie, "Electrowinning and tapping of lanthanum metal,"Bureau of Mines Report of Investigations 6882, U.S. Department of the Interior, Washington, DC, USA, 1966.

48.   R. A. Sharma, "Mettalothermic reduction of rare earth chlorides," US Patent 4680055, General Motors Corporation, http://www. freepatentsonline.com/4680055.pdf.

5

Trace Elements can Influence the Physical Properties of Tooth Enamel

Elnaz Ghadimi[1], Hazem Eimar[1], Benedetto Marelli[2], Showan N Nazhat[2], Masoud Asgharian[3], Hojatollah Vali[1], and Faleh Tamimi[1]

[1]Faculty of Dentistry, McGill University, Montreal QC, Canada

[2]Department of Mining and Materials Engineering, McGill University, Montreal QC, Canada

[3]Department of Mathematics and Statistics, McGill University, Montreal QC, Canada

ABSTRACT

In previous studies, we showed that the size of apatite nanocrystals in tooth enamel can influence its physical properties. This important discovery raised a new question; which factors are regulating the size of these nanocrystals? Trace elements can affect crystallographic properties of synthetic apatite, therefore this study was designed to investigate how trace elements influence enamel's crystallographic properties and ultimately its physical properties.

The concentration of trace elements in tooth enamel was determined for 38 extracted human teeth using inductively coupled plasma-optical emission spectroscopy (ICP-OES). The following trace elements were detected: Al, K, Mg, S, Na, Zn, Si, B, Co, Cr, Cu, Fe, Mn, Mo, Ni, Pb, Sb, Se and Ti. Simple and stepwise multiple regression was used to identify the correlations between trace elements concentration in enamel and its crystallographic structure, hardness, resistance to crack propagation, shade lightness and carbonate content. The presence of some trace elements in enamel was correlated with the size (Pb, Ti, Mn) and lattice parameters (Se, Cr, Ni) of apatite nanocrystals. Some trace elements such as Ti was significantly correlated with tooth crystallographic structure and consequently with hardness and shade lightness. We conclude that the presence of trace elements in enamel could influence its physical properties.

BACKGROUND

Tooth enamel is composed of both an organic and an inorganic phase. The organic phase is composed of proteins such as amelogenin, ameloblastin and tuftelin, as well as minor concentrations of proteoglycans and lipoids (Belcourt and Gillmeth 1979; Eggert et al. 1973; Glimcher et al. 1964). The enamel inorganic phase is composed of well-packed nanocrystals made of calcium phosphate apatite (HA) with small amounts of incorporated trace elements (Sprawson and Bury 1928). The organization and size of apatite crystals in tooth enamel affects its hardness (Jiang et al. 2005) and optical properties (Eimar et al. 20112012). These findings raise the following question: what determines the size of apatite crystals in tooth enamel? One possibility is that the tooth protein content could affect its crystal domain size, however we had found that the concentration of protein in enamel is not associated with the crystallographic structure of mature teeth (Eimar et al. 2012). Therefore this study was designed to investigate other factors, namely the presence of trace elements that can influence the size of apatite crystals in enamel.

The crystallographic properties of synthetic hydroxyapatite (HA) have been found to be influenced by the incorporation of trace elements (Table 1). Some of the trace elements expand the crystal cell lattice parameters of synthetic HA along the a-axis (Fe^{2+}, Fe^{3+}, Sr^{2+} and

Zn^{2+} (molar fraction > 10%)) while others shrink it (SiO_4^{4-}, CO_3^{2-}, Mg^{2+}, Zn^{2+} (molar fraction < 10%) and Ti^{4+}). The crystal domain size along c-axis can be increased by some trace elements (SiO_4^{4-}, CO_3^{2-}, Zn^{2+}, Fe^{2+}, $Fe^{3+}$and Sr^{2+}) and decreased by others (Mg^{2+}, Ni^{2+}, Cr^{3+}, Co^{2+} and Ti^{4+}). Some trace elements can increase the crystallinity (degree of structural order of atoms) and crystal domain size (average length of individual crystals) of synthetic HA (Cr^{3+}, Co^{2+} and Ni^{2+}), while others have the opposite effect (SiO_4^{4-}, Zn^{2+}, CO_3^{2-}, Fe^{2+}, Ti^{4+}, Sr^{2+}, Ce^{3+} and Mg^{2+}) (Christoffersen et al. 1997; Ergun2008; Feng et al. 2005; Hu et al. 2007; Huang et al. 2011; Li et al. 2008; Lin et al. 2007; Mabilleau et al. 2010; Morrissey et al. 2005; Ren et al. 2010; Ribeiro et al. 2006; Tang et al. 2005; Wang et al.2008).

Table 1: Summary of the literature on the effect of trace elements on crystallography parameters in synthetic HA

Crystallographic parameters	Trace elements that increase crystallographic parameters	Trace elements that decrease crystallographic parameters
Lattice along a-axis	Fe^{2+}, Fe^{3+}, Zn^{2+a}, Sr^{2+}	SiO_4^{4-}, CO_3^{2-}, Zn^{2+b}, Ti^{4+}
Lattice along c-axis	Fe^{2+}, SiO_4^{4-}, CO_3^{2-}, Zn^{2+}, Fe^{3+}, Sr^{2+}	Mg^{2+}, Ti^{4+}, Co^{2+}, Ni^{2+}, Cr^{3+}
Crystallinity	Co^{2+}, Ni^{2+}, Cr^{3+}	SiO_4^{4-}, CO_3^{2-}, Zn^{2+}, Fe^{3+}, Ti^{4+}, Mg^{2+}, Ce^{3+}
Crystal domain size along c-axis	Co^{2+}, Ni^{2+}, Cr^{3+}	Fe^{2+}, SiO_4^{4-}, CO_3^{2-}, Zn^{2+}, Ti^{4+}, Mg^{2+}, Ce^{3+}

[a]molar fraction >10%;[b] molar fraction < 10%.

Ghadimi *et al.*

Ghadimi *et al. SpringerPlus* 2013 2:499 doi:10.1186/2193-1801-2-499

Unlike the effect of trace elements on synthetic HA, their role on crystallographic properties of enamel is unknown in the literature. Despite the very low concentration of trace elements in our body, they play a significant role in human body healthiness (Carvalho et al. 1998). Trace elements can enter our body through digestion of food or by exposure to the environment (Lane and Peach 1997) and they can be incorporated into the structure of enamel HA. Trace elements in tooth enamel have been investigated for their role in caries (Curzon and Crocker 1978) and it was found that the presence of F, Al, Fe, Se and Sr is associated with the low risk of tooth caries, while Mn, Cu and Cd have been associated with a high risk (Curzon and Crocker 1978). However, despite their apparent importance on tooth enamel homeostasis, the effect of trace elements on the crystallography and physical properties of enamel remains unknown.

The aim of this study is to find the correlation between the concentration of trace elements detected in tooth enamel and its crystallography and physical-chemical properties. We hypothesized that the incorporation of trace elements in the structure of enamel can affect its crystallography and consequently alter the physical properties of enamel.

RESULTS

Physical Chemical Properties of Tooth Enamel

Among all tooth enamel samples, cell lattice parameters varied along a-axis between 9.40 and 9.47 Å (mean = 9.43 ± 0.004 Å) and along c-axis between 6.84 and 6.92 Å (mean = 6.86 ± 0.004 Å). Unlike the cell lattice parameter along c-axis, the cell lattice parameter along a-axis followed a normal distribution among all samples (Figure 1a, 1b). Crystal domain size along a-axis ranged between 10.31 and 18.08 nm (mean = 13.49 ± 0.349 nm) and along c-axis varied between 18.09 and 25.85 nm (mean = 21.7 ± 0.33 nm) following a normal distribution (Figure 1c, 1d).

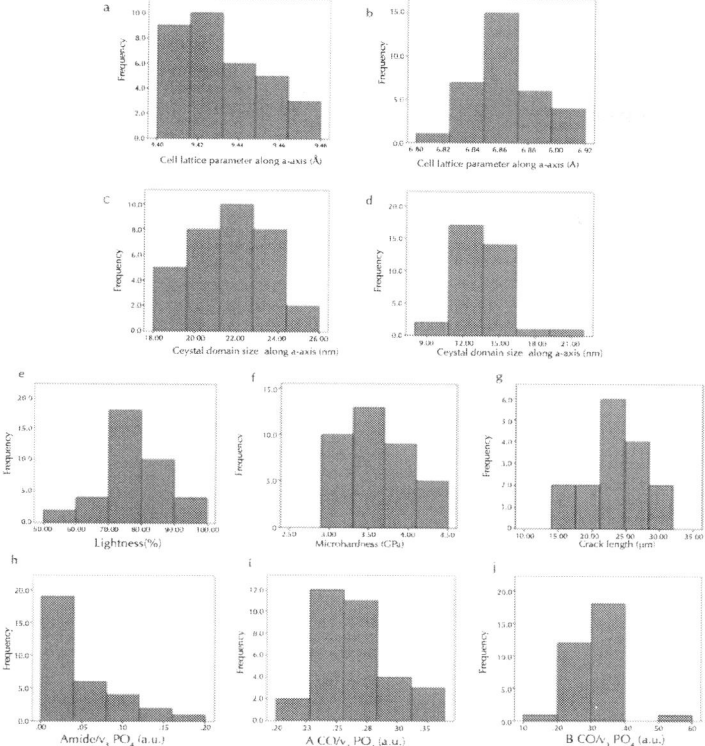

Figure 1: Frequency histograms describing the variations in physical-chemical properties of tooth enamel: cell lattice parameters along (a) a-axis and (b) c-axis; crystal domain length along (c) a-axis and(d) c-axis; (e) tooth shade lightness, (f) hardness and (g) average crack length; (h) the organic relative content and carbonate relative content [(i)type A and (j) type B] in enamel mineral matrix among the examined teeth.

The tooth enamel hardness, crack length and shade lightness followed a normal distribution among the samples analyzed (Figure 1e, 1f and 1g). The hardness values varied between 2.91 and 4.36 GPa (mean = 3.64 ± 0.08 GPa), the crack length varied between 14.60 and 30.02 μm (mean = 23.31 ± 1.49 μm) and the tooth shade lightness values ranged between 59.0 and 96.1% (mean = 79.0 ± 1.8). The relative content of apatite inorganic carbonate type A and type B among the enamel samples followed a normal distribution, while the relative organic content did not (Figure 1h, 1i, 1j).

Trace Elements in Tooth Enamel

A total of 19 trace elements were detected in the tooth enamel samples by ICP. The concentration of the different trace elements varied considerably among tooth enamel samples (Figure 2a). Cr, Mo, Co and Sb had the lowest concentration in tooth enamel compared with other trace elements while Zn, Na and S had the highest one. In order to assess how concentrated were the trace elements in enamel compared to the rest of the body, they were normalized to the average elemental composition of the human body (Frieden 1972; Glover 2003; Zumdahl and Zumdahl 2000). Among the 19 trace elements detected, some of them had a similar concentration in tooth enamel compared to the rest of the body (K and Fe), while others were concentrated by 1 (S, Sb, Pb, Si, Na and Mg), 2 (Mo, Co, Zn, Mn, Cu, Ti and Cr), or 3 orders of magnitude (Se, B, Al and Ni). The largest difference in concentration of trace elements between tooth enamel and the rest of the body belongs to Ni, which Ni is close to 3500 times more abundant in tooth enamel than in the rest of the body.

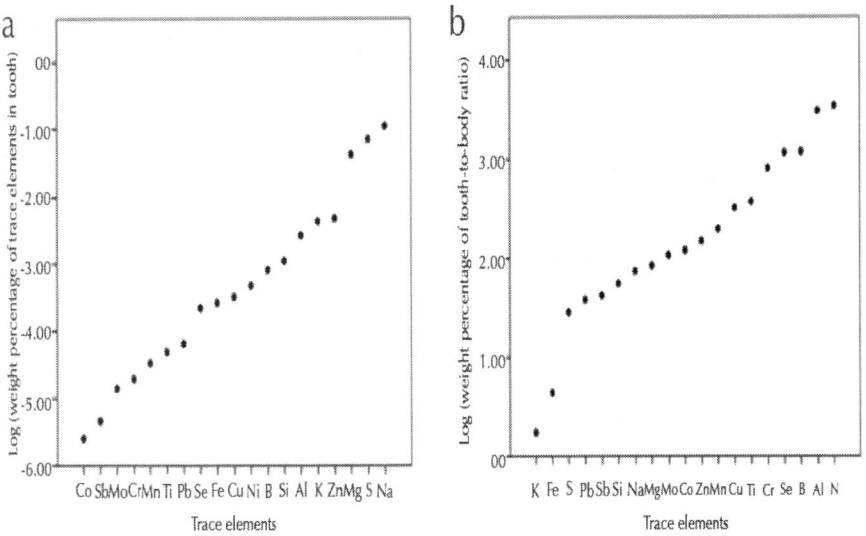

Figure 2: The concentration of trace elements. The figures show (a)the logarithm of trace elements concentration in tooth enamel and (b) the logarithm of tooth-to-body concentration ratio.

The Correlation Analyses of Trace Elements

The simple linear regression was used to find the correlation between each possible pair combination of trace elements in tooth enamel. It was found that the following correlations between trace elements were significant: Al-to-B, Al-to-Sb, Al-to-Si, B-to-Sb, B-to-Si, B-to-Cu, Co-to-Cr, Cr-to-Cu, Cr-to-Ni, Cr-to-Si, Cu-to-Sb, Cu-to-Se, Cu-to-Si, Fe-to-K, Fe-to-Na, Fe-to-Ni, Fe-to-S, Fe-to-Ti, Fe-to-Zn, K-to-Mg, K-to-Mo, K-to-Na, K-to-Ni, K-to-Pb, K-to-S, K-to-Zn, Mo-to-Ni, Na-to-Ni, Na-to-Pb, Na-to-S, Na-to-Zn, Ni-to-S, Pb-to-S, S-to-Zn, Sb-to-Si and Ti-to-Zn. Stepwise multiple regression was done to find the correlation between pairs combination of trace elements adjusting for the presence of other trace elements in tooth enamel. The significant correlations among trace elements concentration using stepwise multiple regression were: Al-to-B, Al-to-Si, B-to-Si, Cu-to-Si, K-to-Mg, K-to-Na, B-to-Cu, Mg-to-Na, Na-to-S, S-to-K, S-to-Mg. The findings showed three independent groups of elements that were directly or indirectly correlated to each other (Figure 3). In the first group, there was a positive correlation among Al-to-B, B-to-Si, B-to-Sb, Si-to-Cu, Cu-to-Se, and a negative correlation between Si and Se. In the second group, Co-to-Cr, Cr-to-Ni, Ni-to-Fe, Fe-to-Ti and Ni-to-Mo were correlated positively; while Ni-to-Ti were correlated negatively. In the third group, all of the correlations (Pb-to-Mg, Mg-to-Na, Na-to-K, K-to-Zn and Na-to-S) were positive. Mn was the only element that was not correlated to any other element.

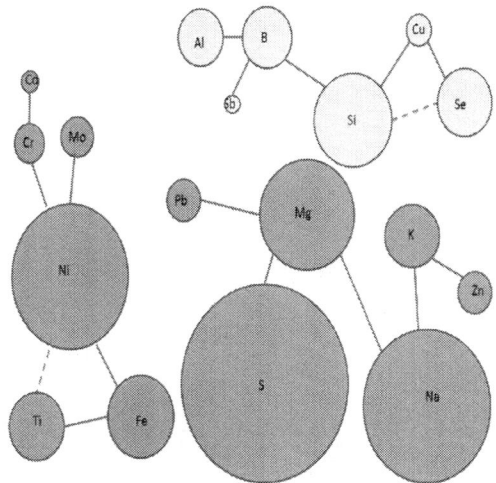

Figure 3: The correlation between trace elements in tooth enamel.In scheme straight lines represent the positive correlations and dotted lines represent the negative ones. Differences in the size of circles are directly proportional to the concentration difference among the correlated elements.

Simple Linear Regression between the Concentration of Elements and Tooth Properties

The correlations between trace element concentration and physical-chemical parameters (crystallographic parameters, hardness, crack length, shade lightness and carbonate content) of enamel samples and the significant ones are summarized in Table 2. There was a strong positive correlation between the concentration of Ti in enamel and tooth hardness, lightness and apatite crystal domain size along c-axis. The correlations between Fe concentration in enamel and both tooth lightness and carbonate type A content in tooth enamel were significant. Also, the correlations between Cu concentration in enamel and both crystal domain size and cell lattice parameter along a-axis were significant. There was a negative association between Al concentration and tooth enamel crack length. A strong inverse relationship was observed between the concentration of Pb and crystal domain size along c-axis. The correlations between the concentration of Se and cell lattice parameter along both a-axis and c-axis were significant. A negative strong correlation was seen between the concentration of Cr and cell lattice parameter along c-axis. Also, there were significant correlations between the concentration of Ni and both, cell lattice parameter along c-axis and carbonate type B. The association between the concentration of S and carbonate type B was significant. The remaining chemical and crystallographic parameters were not correlated to each other.

Table 2: Simple linear and stepwise multiple correlation between trace elements' concentration in enamel and tooth properties

Correlated properties	Correlated elements	Simple linear regression			Multiple stepwise regression		
		R	B	P	R	B	P
Mechanical properties							
Hardness	Ti	0.34	27.00	0.037	0.34	2753.54	0.037
Average crack length	Al	0.67	−18.08	0.005	0.70	−18.08	0.005
Optical properties							
Lightness	Fe	0.46	153.60	0.004	NS	NS	NS
	Ti	0.47	832.79	0.003	0.47	832.79	0.003
Crystallographic structure							
Crystal domain size along a-axis	Cu	0.40	32.87	0.019	NS	NS	NS
Crystal domain size along c-axis	Pb	0.52	−199.11	0.002	0.70	−138.76	0.019
	Ti	0.51	−192.05	0.002		−133.19	0.02
	Mn	NS	NS	NS		184.01	0.025
Cell lattice parameter along a-axis	Cu	0.35	0.24	0.045	NS	NS	NS

		Se	0.59	0.36	<0.001	0.59	0.36	<0.001
Cell lattice parameter along c-axis		Cr	0.38	−6.97	0.031	0.72	−5.79	0.031
		Ni	0.40	−0.09	0.02		−0.06	0.047
		Se	0.52	0.35	0.002		0.36	<0.001
Carbonate Content								
Type A		Fe	0.41	−0.61	0.021	0.41	−0.61	0.021
Type B		Co	0.37	−18.44	0.007	0.77	−16.15	0.011
		Ni	0.70	0.47	<0.001		0.46	<0.001
		S	0.43	0.01	0.014	NS	NS	NS

R: the correlation coefficient; B: the regression coefficient; P: the significance of Pearson correlation and NS: not significant.

Ghadimi *et al.*

Ghadimi *et al.* *SpringerPlus* 2013 2:499 doi:10.1186/2193-1801-2-499

Stepwise Multiple Linear Regression between Elements and Tooth Properties

The results of stepwise multiple regression between the concentration of trace elements in enamel and the mechanical properties, optical properies, crystallographic structure and carbonate content of the teeth are presented in Table 2. Most of the results from the stepwise multiple regression analysis confirmed the results of the simple linear regression. It was found that the concentration of Ti was associated with tooth enamel hardness, lightness and crystal domain size along c-axis. The concentration of Pb and Ti in enamel had a negative correlation with tooth enamel crystal domain size while Mn had a positive one. The correlation between the concentration of Al and crack length in tooth enamel was negative. The concentration of Se was associated positively with the cell lattice parameters along both a-axis and c-axis. The concentration of Cr and Ni had an inverse correlation with the cell lattice parameter along c-axis. The concentrations of Fe and Co were negatively correlated to the carbonate type A and type B, respectively. The concentration of Ni and carbonate type B were associated positively.

The protein content in tooth enamel was correlated to none of the trace elements. Some of the trace elements such as Fe, S and Cu were correlated to the enamel properties by stepwise multiple regression but not by simple linear regression and vice versa (such as Mn).

DISCUSSION

In this study 12 trace elements were found to be associated with tooth composition, structure and physical properties. Underneath we compared our findings with previous studies and we discussed the possible sources of these contaminants.

Trace Elements in Tooth Enamel

Trace elements enter the human body coming from different sources such as food, water, air, etc. (Kampa and Castanas 2008; Malczewska–Toth 2012). They can be incorporated in tooth enamel structure and in this study we show that they could affect the physical chemical properties of enamel. Below the association between each detected trace elements in tooth enamel with the tooth physical and chemical properties are detailed.

Selenium

Selenium is a non-metallic element widely distributed in nature that can be absorbed by the body through oral intake or breathing (Malczewska–Toth 2012). Selenium is incorporated in synthetic HA through anionic exchange of phosphate with selenite in a one-to-one (1:1) substitution ratio (Monteil-Monteil-Rivera et al. 2000). In our study, we found that the presence of Se in enamel was associated with increased lattice parameters along a-axis and c-axis. This was logically expected because the ionic radius of Se^{4+} (0.50 Å) is larger than the ionic radius of P^{5+} (0.35 Å), so by substitution of Se in synthetic HA, the lattice parameters increase (Ma et al. 2013). However ours is the first study that reports this phenomenon in tooth enamel.

Chromium

Chromium is a heavy metal that is essential for the body in small amounts (Kampa and Castanas2008). Its main role is in controlling the fat and sugar metabolism (Kimura 1996), and it helps to increase the muscle and tissue growth (Schroeder et al. 1963). It can enter human body through water, air and food (Kampa and Castanas 2008).

Although the ionic radius of Ca^{2+} (0.99 Å) is much bigger than Cr^{3+} (0.69 Å), Cr can exchange with Ca in synthetic HA (Chantawong et al. 2003). Accordingly, the substitution of Cr^{3+} in synthetic HA decreases the cell lattice parameters along both a-axis and c-axis, which might explain why we observed a significant association between the concentration of Cr and cell lattice parameter along c-axis in tooth enamel (Mabilleau et al. 2010).

Nickel

Nickel is a toxic metal that can be absorbed by the body through water, air or food (Kampa and Castanas 2008). Ni is incorporated in HA through substitution of $Ca^{2+}(I)$ and bonds with O to form Ni_3PO_4 (Zhang et al. 2010). The ionic radius of Ca^{2+} (0.99 Å) is much bigger than Ni^{2+} (0.72 Å) (Chantawong et al. 2003). Consequently, with addition of Ni^{2+}, the cell lattice parameter along c-axis decreases in synthetic HA (Mabilleau et al. 2010). These observations are in agreement with our study which to best of our knowledge is the first to report the inverse association between the concentration of Ni and crystal domain size in tooth enamel. Also, we found that the substitution of Ni in tooth enamel had a strong positive association with the presence of carbonate type B. Future studies will have to be performed in order to understand this phenomenon.

Cobalt

Cobalt is a toxic metal that exists in the environment and can enter the body through water, air and food (Barceloux and Barceloux 1999; Duruibe et al. 2007). In HA, Ca^{2+} is substituted by Co^{2+} following the equation Ca_{10-x} Co_x $(PO_4)_6$ $(OH)_2$ (Elkabouss et al. 2004) and the maximum exchange of Co with Ca is 1.35 wt% Co (Elkabouss et al.

2004). In this study, we report for the first time that the incorporation of Co in tooth enamel structure had a strong negative association with the substitution of carbonate type B. More studies are required to find the reason behind this phenomenon.

Lead

Lead is a poisonous heavy metal that can harm the human body (Shukla and Singhal 1984). It can enter the body through water, air or food (Kampa and Castanas 2008). In HA, at low concentrations, Pb^{2+} ions (1.2 Å) replace Ca^{2+} ions (0.99 Å) (Mavropoulos et al. 2002; Miyake et al. 1986; Prasad et al. 2001) at the calcium site II following the equation $Pb_{(10-x)}Ca_x(PO_4)_6(OH)_2$ (Mavropoulos et al.2002). Accordingly the sorption of low concentrations of lead by synthetic HA decreases its crystal domain size (Mavropoulos et al. 2002) and this could be the reason why we found in our study that the presence of lead had a negative correlation with the size of enamel apatite crystal.

Titanium

Titanium is a metal commonly used in the field of biomaterials and bio-applications (Niinomi 2002). Titanium characteristics such as high strength, low modulus of elasticity, low density, biocompatibility, complete inertness to body environment and high capacity to integrate with bone and other tissues make it widely used in implant applications (Niinomi 2002). Ti ions are absorbed by body through food like candies, sweets and chewing gums (Weir et al. 2012).

The ionic radius of Ti^{4+} (0.68 Å) is much smaller than the ionic radius of Ca^{2+} (0.99 Å) (Ribeiro et al.2006), so the substitution of Ca by low concentrations of Ti in synthetic HA results in decreased cell lattice parameters and crystal domain size (Ergun 2008). We found that the concentration of Ti in tooth enamel HA was associated with decreased enamel crystal domain size, which is in agreement with the previous studies on synthetic HA (Ergun 2008; Hu et al. 2007; Huang et al. 2010).

In this study, the presence of Ti in tooth enamel was associated with increasing tooth hardness. Since Ti had an inverse association with the crystal domain size and the size of apatite nanocrystals in tooth enamel is inversely correlated to tooth hardness (Eimar et al. 2012), this could be the reason behind the positive association between Ti concentration in enamel apatite and hardness.

We previously showed that the tooth enamel crystal domain size was associated with its optical properties; when the enamel crystal domain size is larger, its lightness is lower (Eimar et al. 2011). This phenomenon is due to the fact that more light can be scattered from tooth enamel composed of small crystals (Eimar et al. 2011). In this study we found that the presence of Ti in tooth enamel structure was associated with both smaller crystal domain size and higher lightness, which confirmed our previous observation (Eimar et al. 2011).

Manganese

Manganese is another trace element that can be uptaken from food, air and water (Frieden 1984; Kampa and Castanas 2008). Mn^{2+} replaces Ca^{2+} in HA (Medvecky et al. 2006). Previous studies have shown that the incorporation of Mn^{2+} in synthetic HA does not change the crystal domain size significantly (Medvecky et al. 2006; Ramesh et al. 2007). In our study we showed that the presence of Mn in tooth enamel was associated with its apatite crystal domain size. Further studies are needed to understand this phenomenon.

Iron

Iron is an essential element for human life, it can enter the body through food such as vegetables and it is one of the trace elements found in teeth (Cook et al. 1972). Fe affects the carbonate content in synthetic HA (Low et al. 2010). It was found that in low concentration of Fe, Carbonate type A can be substituted by Fe in synthetic HA (Low et al. 2010). We found that the incorporation of Fe in tooth enamel had lower relative content of carbonate type A. Further studies are needed to understand this phenomenon.

Aluminum

Aluminum is one of the elements found in the body that can be absorbed through air, food, and water (Campbell et al. 1957; Maienthal and Taylor

1968; Oke 1964). Its concentration in the body increases with age and at higher amounts can cause brain and skeleton disorders (Alfrey et al. 1976; Little and Steadman 1966). Also, the discoloration of tooth enamel can be seen in the presence of Al (Little and Steadman 1966). In this study, we found that the concentration of Al was correlated negatively with crack length in tooth enamel. Teeth with lower level of Al were more prone to have longer cracks and vice versa. More studies are needed to understand this phenomenon.

Sources of Trace Elements

Trace elements are distributed differently among tooth enamel and dentin. For example Cu, Pb, Co, Al, I, Sr, Se, Ni and Mn are more abundant in tooth enamel compared to dentine while Fe and F are more concentrated in dentine than enamel (Derise and Ritchey 1974; Lappalainen and Knuuttila1981). Also trace elements vary among the different layers of enamel. Fe, Pb and Mn are more abundant in the outer layers of enamel compared to the inner ones (Brudevold and Steadman 1956; Reitznerová et al. 2000). These findings seem to indicate that certain trace elements could be coming from the environment (i.e. Mn and Fe) and are incorporated after eruption or deposed in tooth enamel during calcification (Brudevold et al. 1975; Nixon et al. 1966; Okazaki et al. 1986). Trace elements can enter the human body coming from different sources such as food, water and air (Cleymaet et al. 1991; Cook et al. 1972; Duruibe et al. 2007). Underneath we discuss the dental products and fluids (i.e. saliva, dental prosthesis and dental porcelain) as possible sources of trace elements in tooth enamel.

Saliva

Teeth are washed constantly by saliva (Duggal et al. 1991). Several trace elements found in saliva are known to affect the composition of enamel surface (Reitznerová et al. 2000). The concentration of trace elements in saliva varies among people. These are some of the elemental detection in enamel with the range of Cu (20–321 μgL^{-1}), Zn (28–358 μgL^{-1}), Mg (2.53-34.13 mgL^{-1}), Ca (21.5-170 mgL^{-1}), Al (25–102 μgL^{-1}), Sr (9–26 μgL^{-1}), Mn (2.3-5 μgL^{-1}), Fe (25–160 μgL^{-1}), Si

(2–3 µgL^{-1}), Na (257.49-276.8 mg/L), K (958.73-994.704 mg/L) and Cr (0.8-3.6 µgL^{-1}) (Borella et al. 1994; Grad 1954; Sighinolfi et al. 1989).

The most abundant trace elements in saliva (Na, Mg, K and Zn) are also the most abundant trace elements in tooth enamel (Borella et al. 1994; Grad 1954; Sighinolfi et al. 1989). Interestingly, in our study the concentration of each one of these trace elements was directly or indirectly correlated to each other. These observations indicated that saliva could influence the composition of enamel.

Dental Prosthesis

Partial denture could be a possible source of trace elements found in teeth. The alloys commonly used to fabricate dental prosthesis include Cr, Co, Ni, Fe, Ti and Mo (Andersson et al. 1989; Asgar et al.1970; Morris et al. 1979; Yamauchi et al. 1988). The concentration of Cr, Co, Fe and Ni in saliva of patients with partial dentures is higher than in patients without partial dentures (de-Melo et al. 1983; Gjerdet et al. 1991).

The concentration of Cr, Co, Mo, Ni, Ti and Fe in tooth enamel was correlated strongly to each other (Figure 3). This finding along with the fact that these materials are found in the composition of dental prosthesis seems to indicate that the source of these metals in enamel could be dental prosthesis. Therefore, the presence of denture in mouth can affect the concentration of trace elements in tooth enamel. Future studies will be performed to confirm the effect of trace elements found in dental prosthesis in tooth enamel.

Dental Porcelain

Dental porcelain is composed of a leucite crystallite phase and a glass matrix phase (Panzera and Kaiser 1999) and its elemental composition includes Si (57-66%), B (15-25%), Al (7-15%), Na (7-12%), K (7-15%) and Li (0.5-3%) (Panzera and Kaiser 1999; Sekino et al. 2001). Interestingly, the concentration of these elements in tooth enamel was strongly correlated to each other (Figure 3). This result seems to indicate that the dental porcelain might be another possible source of these elements in tooth enamel. Although future studies will have to be performed to confirm this possibility.

Clinical Implications

Trace elements can enter the structure of tooth enamel and affect its physical-chemical properties. In this study, we found that there are several sources for trace elements to enter enamel structure such as saliva, dental prosthesis or dental porcelain. Future studies will have to be performed to determine the effect of saliva and dental prosthesis on tooth enamel structure.

Metallic components of dental prosthesis are usually based on Cr, Co and Ni. These metals can cause sensitivities and allergies (Blanco-Dalmau et al. 1984; Brendlinger and Tarsitano 1970) and for these reasons Ti based dentures have been developed (Andersson et al. 1989; Yamauchi et al. 1988). In our study, we confirmed that the presence of Ti in tooth enamel could be beneficial by rendering teeth whiter and harder. Therefore, dental materials containing Ti could have additional benefits, besides the ones that already are known and used in biomaterial sciences.

Limitations

One limitation of the present study is relatively small sample size and another one is the limited number of detected trace elements in each tooth sample. With increasing the number of sample size or using another technique in order to promote the detection limit, we might find more trace elements in the samples and more significant correlations between the concentration of trace elements and the physical-chemical properties of it.

CONCLUSIONS

The presence of trace elements in tooth enamel could influence the physical chemical properties of tooth enamel. In this study we found that the concentration of Ti in tooth enamel had a strong relationship with enamel hardness, lightness and crystal domain size along c-axis. The incorporation of Fe had a negative association with the presence of carbonate type A, while the incorporation of Co and Ni was correlated with the formation of carbonate type B. The concentration of Al in tooth enamel was inversely correlated with the length of cracks forming in enamel. Also, we found that the presence of Se in tooth enamel had a positive correlation with cell lattice parameters along both a-axis and c-axis while Cr and Ni had a negative correlation with cell lattice parameters along c-axis. Also, it was shown that the concentration of Pb and Mn in tooth enamel had a positive association with tooth enamel crystal domain size along c-axis.

MATERIALS AND METHODS

After obtaining ethical approval from McGill University Health Center (MUHC) ethical committee, a set of 38 extracted human teeth were collected from patients attending McGill Undergraduate Dental Clinic, scheduled for extractions. In this study, the included teeth were sound human upper anterior teeth. Teeth with caries, demineralized areas, cracks, cavitations, restorations, severe or atypical intrinsic stains, and/ or tooth bleaching history were excluded.

Upon extraction, teeth were immersed in 10% formalin solution (BF-FORM, Fisher Scientific, Montreal, Canada) for 1 week, before cleaning them in an ultrasound bath (FS$_2$0D Ultrasonic, Fisher Scientific, Montreal, Canada) with de-ionized distilled water at 25°C for 60 minutes. Then, they were polished with a low-speed dental handpiece (M5Pa, KAB-Dental, Sterling Heights, MI) using SiC cups (Pro-Cup, sdsKerr, Orange, CA) and dental prophylaxis pumice of low abrasive capability (CPR™, ICCARE, Irvine, CA) for 1 minute. Teeth were rinsed again in an ultrasonic bath with de-ionized distilled water before storing them in labeled Eppendorf tubes with a 10% formalin solution.

Tooth Spectrophotometry

Tooth shade was registered by tooth spectrophotometry (Easy shade®, Vita Zahnfabrik, Germany) which is the most accurate and reproducible technique used for tooth shade masurements (Chu et al.2010; Paul et al. 2002). Shade measurements were collected using the parameters of Munsell's colour system (L*C*H*) and they were repeated three times for each tooth. The mean and the standard deviation for each shade parameter were calculated. Tooth dehydration might induce changes in tooth shade so, in order to avoid its dehydration, during shade measurement each tooth was kept wet at all times.

Vickers Microhardness

A sagittal section was obtained from each tooth using a carbide bur (FG56, sds Kerr, Orange, CA) adapted to a high-speed dental handpiece (TA-98LW, Synea, Bürmoos, Austria) and cooled with de-ionized distilled water in order to prevent overheating. Each tooth section was fixed in clear methylmethacrylate resin (DP-Ortho-F, DenPlus, Montreal, QC). The resulting blocks were mirror polished using ascending grits of silicon carbide papers with de-ionized distilled water (Paper-c wt, AAAbrasives, Philadelphia, PA) (240, 400, 600, 800 and 1200) and were smoothed with a polishing cloth.

A Vickers microhardness device (Clark CM100 AT, HT-CM-95605, Shawnee Mission, KS) was used to make indentations on the polished surfaces of tooth enamel. The indentation load was 300 N with a loading time of 10 seconds. Due to the variation in microhardness values within each tooth enamel sample, six indentations were performed between the DEJ and external surface of each enamel sample (Bembey et al. 2005; Gutiérrez-Salazar and Reyes-Gasga 2003; Newbrun and Pigman 1960; White et al. 2000). A minimum distance of 50μm was maintained, between the successive indentations. A computer software (Clemex Vision PE 3.5, Clemex Technologies Inc, Shawnee Mission, KS) was used to measure the microhardness value at the site of indentation from images captured with a built-in camera.

Enamel Crack Propagation

Indentations were made on the polished surfaces of tooth enamel that prepared as described above, with a Vickers microhardness device (Clark CM100 AT, HT-CM-95605, Shawnee Mission, KS). Half way between the DEJ and the surface of enamel 7 indentations were applied on each tooth enamel sample. Between the successive indentations, a minimum distance of 250μm was maintained. Indentation load was 500N with a loading time of 10 seconds. Upon indentation, cracks emanated from the corners of each indentation. Then, the samples were sputter-coated with gold and images of the cracks were captured with a VP-SEM (Hitachi S -3000N VP, Japan) at 500 magnification (Figure 4c). The length of the cracks was measured using the ImageJ software (US National Institutes of Health, Bethesda, MD). The average crack length for each indentation was calculated by summing up the length of cracks and dividing by the number of cracks (Chicot et al. 2009; Roman et al. 2002).

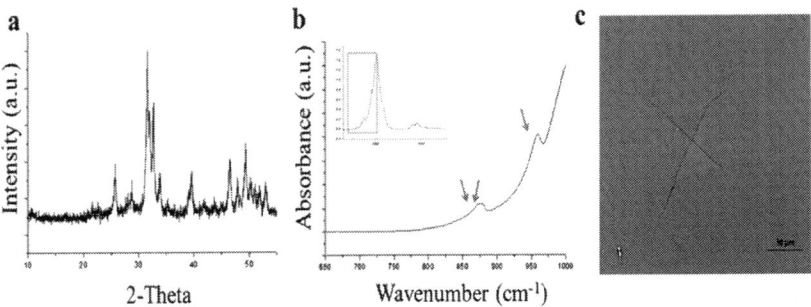

Figure 4: XRD, FTIR spectrums and SEM micrograph of Vicker's microindentation on tooth enamel samples. (a) XRD spectrum of tooth enamel powder, (b) FTIR absorbance spectra of tooth enamel samples normalized to absorbance peak of $_3PO_4$ at 1013 cm^{-1} (arrows show the peaks that used for the calculation of carbonate content), (c) SEM micrograph of Vicker's microindentation on enamel (cracks are shown by arrows on the corners of indentation).

X-Ray Powder Diffraction

To determine the crystallographic dimension of apatite in each tooth enamel sample, X-ray powder diffraction (XRD) (D8-Discover/GADDS, Bruker, Karlsruhe, Germany) was used. The XRD patterns were recorded using a diffractometer with CuK radiation (setting: 40 Kv, 40 mA, 10-60° scanning angle, 0.02 step size and 1800 scan step time) (Hanlie et al. 2006; Xue et al. 2008). DIFFRAC-plus EVA software (AXS, Bruker, Karlsruhe, Germany) was used to analyze the data obtained from each XRD spectrum (Figure 4a) (Xue et al. 2008).

The average HA crystal domain size along c-axis and a-axis for each enamel sample were calculated using the (002) and (310) Bragg peaks of the XRD spectrum and Scherrer's formula (Eq. 1)

$$D = \frac{k\lambda}{\beta \cos}$$

(1)

where D is the average of domain lengths, K is the shape factor, is the x-ray wavelength is the line broadening at half the maximum intensity (FWHM) and is the Bragg angle.

The enamel crystal cell lattice parameters, a-axis and c-axis were calculated using the XRD (002) and (310) Bragg peaks relying on the following equation (Hanlie et al. 2006),

$$\frac{1}{d^2} = \left(\frac{4}{3}\right) * \left(\frac{h^2 + hk + k^2}{a^2}\right) + \left(\frac{l^2}{c^2}\right)$$

(2)

where d is the spacing between adjacent planes (interplanar spacing) in the crystal, hkl are the miller indices that are the reciprocal intercepts of the plane on the unit cell axes, a is the a-axis and c is the c-axis. We used the (002) and (310) Bragg peaks for our crystallography calculation, because they have been widely used in the literature and they do not overlap with other peaks (Hanlie et al. 2006; Leventouri et al. 2009; Simmons et al. 2011).

FTIR

The chemical composition of enamel was investigated by FTIR spectroscopy (Spectrum 400, Perkin–Elmer, Waltham, MA). The enamel powder samples were submitted to an FTIR spectrophotometer using a single bounce ZnSe diamond-coated ATR crystal. For each sample, a total of 64 scans per run at 2 cm^{-1} resolution were used (Figure 4b). FTIR studies were carried out in the range 700–1800 cm^{-1}. The collected spectra were normalized according to the absorbance of v_3PO_4 at 1013cm^{-1} using the FTIR spectrophotometry software (Spectrum, Perkin-Elmer, USA).

According to previous studies, the organic content of enamel was estimated from the Amide I-to-v_3PO_4 ratio (Aparicio et al. 2002; Bartlett et al. 2004; Bohic et al. 2000). The carbonate content within enamel mineral matrix was estimated from the ratios of v_2CO_3 type A (~878cm^{-}

^{1}) and B ($\sim872cm^{-1}$) to the v_3PO_4 and v_1PO_4 ($\sim960cm^{-1}$) absorption bands (Antonakos et al. 2007; Lasch et al. 2002; Rey et al. 1989).

Inductively Coupled Plasma-Optical Emission Spectroscopy

The concentration of trace elements in tooth enamel samples was determined with Inductively Coupled Plasma-Optical Emission Spectroscopy (ICP-OES) (Thermo Scientific iCAP 6500, Cambridge, UK). Weighed tooth enamel powdered samples were dissolved in concentrated nitric acid (5 ml; 68% wt/wt) at a temperature of 95°. Yttrium (5 ppm) was added to the solution as an internal standard to make corrections for possible sample preparation errors and sample matrix corrections. After 2 hours of acid digestion, the liquids of the resulting solutions (0.25 ml) were diluted into both deionized-distilled water (10 ml) and 4% nitric acid (25 ml), separately. Both diluted solutions were submitted to ICP-OES using the following setup: power of 1150W, auxiliary gas-flow rate of 0.5 L/min, nebulizer gas-flow rate of 0.5 L/min, sample flow rate of 0.7 ml/min, cooling gas of 12 L/min and integration time of 10 sec. The operating software ITEVA (version 8) was used to control the instrument function and data handling. The quality control checks were done prior to initial analysis and every 12 consecutive samples.

Data Analysis

The correlations between the concentration of each trace element in tooth enamel and its physical chemical properties were determined using simple linear regression. Due to high correlations between trace elements in tooth enamel and their possible effect on the results, in addition to simple linear regression, stepwise multiple regression was performed. Stepwise multiple regression is more accurate than simple linear regression, because it provides us information about the correlation of each trace element with others while adjusting for their inter-correlations. Also, the association of trace elements with physical-chemical properties of tooth enamel were obtained using the stepwise multiple regression, adjusting for the inter-correlation between trace elements. The statistical significance was set at $P < 0.05$

and all statistical analyses were done using SPSS 19 software (IBM, New York, NY).

AUTHORS' CONTRIBUTIONS

FT planned the research. HV and FT supervised the research. EG and HE prepared the samples. EG and FT drafted the manuscripts. BM and SN aided in running FTIR. EG and MA performed the statistical analysis. All authors read and approved the final manuscript.

ACKNOWLEDGEMENTS

The authors would like to acknowledge the "Fondation de l'Ordre des dentistes du Québec" (FODQ), the Faculty of Dentistry of McGill University and the Natural Sciences and Engineering Research Council of Canada (NSERC-Discovery; F.T.) for their financial support.

REFERENCES

1. Alfrey AC, LeGendre GR, Kaehny WD (1976) The dialysis encephalopathy syndrome. New Engl J Med 294:184-188

2. Andersson M, Bergman B, Bessing C, Ericson G, Lundquist P, Nilson H (1989) Clinical results with titanium crowns fabricated with machine duplication and spark erosion. Acta Odontol 47:279-286

3. Antonakos A, Liarokapis E, Leventouri T (2007) Micro-Raman and FTIR studies of synthetic and natural apatites. Biomaterials 28:3043-3054

4. Aparicio S, Doty S, Camacho N, et al. (2002) Optimal methods for processing mineralized tissues for Fourier transform infrared microspectroscopy. Calcif tissue int 70(5):422-429

5. Asgar K, Techow BO, Jacobson JM (1970) A new alloy for partial dentures. J Prosthet Dent 23:36-43PubMed Abstract

6. Barceloux DG, Barceloux D (1999) Cobalt. Clin Toxic 37:201-216

7. Bartlett J, Beniash E, Lee D, Smith C (2004) Decreased mineral content in MMP-20 null mouse enamel is prominent during the maturation stage. J Dent Res 83:909-913

8. Belcourt A, Gillmeth S (1979) EDTA soluble protein of human mature normal enamel. Calcif Tissue Int 28:227-231

9. Bembey AK, Oyen ML, Ko C, Bushby AJ, Boyde A (2005) Elastic modulus and mineral density of dentine and enamel in natural caries lesions. Mater Res Soc Symp Proc 874:125

10. Blanco-Dalmau L, Carrasquillo-Alberty H, Silva-Parra J (1984) A study of nickel allergy. J Prosthet Dent 52:116-119

11. Bohic S, Rey C, Legrand A, et al. (2000) Characterization of the trabecular rat bone mineral: effect of ovariectomy and bisphosphonate treatment. Bone 26:341-348

12. Borella P, Fantuzzi G, Aggazzotti G (1994) Trace elements in saliva and dental caries in young adults. Sci Total Environ 153:219-224

13. Brendlinger DL, Tarsitano J (1970) Generalized dermatitis due to sensitivity to a chrome cobalt removable partial denture. J Am Dent Assoc 81:392

14. Brudevold F, Steadman LT (1956) The distribution of lead in human enamel. J Dent Res 35:430-437

15. Brudevold F, Reda A, Aasenden R, Bakhos Y (1975) Determination of trace elements in surface enamel of human teeth by a new biopsy procedure. Arch Oral Biol 20:667-673

16. Campbell IR, Cass J, Cholak J, Kehoe R (1957) Aluminum in the environment of man; a review of its hygienic status. AMA Arch Ind Health 15:359 PubMed Abstract

17. Carvalho ML, Brito J, Barreiros MA (1998) Study of trace element concentrations in human tissues by EDXRF spectrometry. X-Ray Spectrometry 27:198-204

18. Chantawong V, Harvey NW, Bashkin VN (2003) Comparison of heavy metal adsorptions by thai kaolin and ballclay. Water Air and Soil Poll 148:111-125

19. Chicot D, Duarte G, Tricoteaux A, Jorgowski B, Leriche A, Lesage J (2009) Vickers Indentation Fracture (VIF) modeling to analyze multi-cracking toughness of titania, alumina and zirconia plasma sprayed coatings. Mater Sci Eng A 527:65-76

20. Christoffersen J, Christoffersen MR, Kolthoff N, Barenholdt O (1997) Effects of strontium ions on growth and dissolution of hydroxyapatite and on bone mineral detection. Bone 20:47-54

21. Chu SJ, Trushkowsky RD, Paravina RD (2010) Dental color matching instruments and systems. Review of clinical and research aspects. J Dent 38:e2

22. Cleymaet R, Bottenberg P, Slop D, Clara R, Coomans D (1991) Study of lead and cadmium content of surface enamel of schoolchildren from an industrial area in Belgium. Community Dent Oral Epidemiol 19:107-111

23. Cook J, Layrisse M, Martinez-Torres C, Walker R, Monsen E, Finch C (1972) Food iron absorption measured by an extrinsic tag. J Clin Invest 51:805 |PubMed

24. Curzon ME, Crocker DC (1978) Relationships of trace elements in human tooth enamel to dental caries. Arch Oral Biol 23:647-653

25. De-Melo JF, Gjerdet NR, Erichsen ES (1983) Metal release from cobalt-chromium partial dentures in the mouth. Acta Odontol Scand 41:71-74

26. Derise NL, Ritchey S (1974) Mineral composition of normal human enamel and dentin and the relation of composition to dental caries: II. Microminerals. J Dent Res 53:853-858

27. Duggal M, Chawla H, Curzon M (1991) A study of the relationship between trace elements in saliva and dental caries in children. Arch Oral Biol 36:881-884

28. Duruibe J, Ogwuegbu M, Egwurugwu J (2007) Heavy metal pollution and human biotoxic effects. Int J Phys Sci 2:112-118

29. Eggert FM, Allen GA, Burgess RC (1973) Amelogenins. Purification and partial characterization of proteins from developing bovine dental enamel. Biochem 131:471-484

30. Eimar H, Marelli B, Nazhat SN, et al. (2011) The role of enamel crystallography on tooth shade. J Dent 39(Suppl 3):e3-e10.

31. Eimar H, Ghadimi E, Marelli B, et al. (2012) Regulation of enamel hardness by its crystallographic dimensions. Acta biomater 8:3400-3410.

32. Elkabouss K, Kacimi M, Ziyad M, Ammar S, Bozon-Verduraz F (2004) Cobalt-exchanged hydroxyapatite catalysts: magnetic studies, spectroscopic investigations, performance in 2-butanol and ethane oxidative dehydrogenations. J Catal 226:16-24

33. Ergun C (2008) Effect of Ti ion substitution on the structure of hydroxylapatite. J Eur Ceram Soc 28:2137-2149

34. Feng Z, Liao Y, Ye M (2005) Synthesis and structure of cerium-substituted hydroxyapatite. J Mater Sci Mater Medicine 16:417-421

35. Frieden E (1972) The chemical elements of life. Sci Am 227:52-60 PubMed Abstract.

36. Frieden E (1984) Biochemistry of the essential ultratrace elements. Plenum New York 3:89-132

37. Gjerdet NR, Erichsen ES, Remlo HE, Evjen G (1991) Nickel and iron in saliva of patients with fixed orthodontic appliances. Acta Odontol 49:73-78

38. Glimcher MJ, Friberg UA, Levine PT (1964) The isolation and amino acid composition of the enamel proteins of erupted bovine teeth. Biochem J 93:202-10

39. Glover TJ (2003) Pocket ref. Littleton, Colo: Sequoia Pub.

40. Grad B (1954) Diurnal, age, and sex changes in the sodium and potassium concentration of human saliva. J gerontol 9:276-286

41. Gutiérrez-Salazar MP, Reyes-Gasga J (2003) Microhardness and chemical composition of human tooth. Mater Res 6:367-373

42. Hanlie H, Liyun T, Tao J (2006) The crystal characteristics of enamel and dentin by XRD method. J Wuhan Univ Technol 21:9-12

43. Hu AM, Li M, Chang CK, Mao DL (2007) Preparation and characterization of a titanium-substituted hydroxyapatite photocatalyst. J Mol Catal A-Chem 267:79-85

44. Huang J, Best SM, Bonfield W, Buckland T (2010) Development and characterization of titanium-containing hydroxyapatite for medical applications. Acta biomater 6:241-9

45. Huang T, Xiao YF, Wang SL, et al. (2011) Nanostructured Si, Mg, CO3 (2-) substituted hydroxyapatite coatings deposited by liquid precursor plasma spraying: synthesis and characterization. J Therm Spray Technol 20:829-836

46. Jiang H, Liu XY, Lim CT, Hsu CY (2005) Ordering of self-assembled nanobiominerals in correlation to mechanical properties of hard tissues. Appl Phys Lett 86:163901163901-3.

47. Kampa M, Castanas E (2008) Human health effects of air pollution. Environ Pollut 151:362-367

48. Kimura K (1996) [Role of essential trace elements in the disturbance of carbohydrate metabolism]. Nihon rinsho JPN J. Clin Med 54:79

49. Lane DW, Peach DF (1997) Some observations on the trace element concentrations in human dental enamel. Biol Trace Elem Res 60:1-11

50. Lappalainen R, Knuuttila M (1981) The concentrations of Pb, Cu, Co and Ni in extracted permanent teeth related to Donors' Age and elements in the soil. Acta Odontol Scand 39:163-167

51. Lasch P, Pacifico A, Diem M (2002) Spatially resolved IR microspectroscopy of single cells. Biopolymers 67:335-338

52. Leventouri T, Antonakos A, Kyriacou A, Venturelli R, Liarokapis E, Perdikatsis V (2009) Crystal structure studies of human dental apatite as a function of age. Int J Biomater 2009.

53. Li M, Xiao X, Liu R, Chen C, Huang L (2008) Structural characterization of zinc-substituted hydroxyapatite prepared by hydrothermal method. J Mater Sci Mater Med 19:797-803

54. Lin YG, Yang ZR, Jiang C (2007) Preparation, characterization and antibacterial property of cerium substituted hydroxyapatite nanoparticles. J Rare Earth 25:452-456

55. Little M, Steadman L (1966) Chemical and physical properties of altered and sound enamel—IV: Trace element composition. Arch Oral Biol 11:273-IN1.

56. Low H, Ritter C, White T (2010) Crystal structure refinements of the 2H and 2M pseudomorphs of ferric carbonate-hydroxyapatite. Dalton Trans 39:6488-6495

57. Ma J, Wang Y, Zhou L, Zhang S (2013) Preparation and characterization of selenite substituted hydroxyapatite. Mater Sci Eng C 33:440-445

58. Mabilleau G, Filmon R, Petrov PK, Basle MF, Sabokbar A, Chappard D (2010) Cobalt, chromium and nickel affect hydroxyapatite crystal growth in vitro. Acta Biomater 6:1555-1560

59. Maienthal EJ, Taylor JK (1968) Polarographic methods in determination of trace inorganics in water. Trace Inorg Water Adv Chem 73:172-182

60. Malczewska–Toth B (2012) Phosphorus, Selenium, Tellurium, and Sulfur. John Wiley and sons. pp 841-884

61. Mavropoulos E, Rossi AM, Costa AM, Perez CA, Moreira JC, Saldanha M (2002) Studies on the mechanisms of lead immobilization by hydroxyapatite. Environ Sci Technol 36:1625-9

62. Medvecky L, Stulajterova R, Parilak L, Trpcevska J, Durisin J, Barinov SM (2006) Influence of manganese on stability and particle growth of hydroxyapatite in simulated body fluid. Colloids Surf A 281:221-229

63. Miyake M, Ishigaki K, Suzuki T (1986) Structure refinements of Pb2+ ion-exchanged apatites by x-ray powder pattern-fitting. J Solid State Chem 61:230-235

64. Monteil-Rivera F, Fedoroff M, Jeanjean J, Minel L, Barthes MG, Dumonceau J (2000) Sorption of Selenite (SeO(3)(2-)) on Hydroxyapatite: An Exchange Process. J Colloid Interface Sci 221:291-300

65. Morris HF, Asgar K, Rowe AP, Nasjleti CE (1979) The influence of heat treatments on several types of base-metal removable partial denture alloys. J Prosthet Dent 41:388-395

66. Morrissey R, Rodriguez-Lorenzo LM, Gross KA (2005) Influence of ferrous iron incorporation on the structure of hydroxyapatite. J Mater Sci Mater Med 16:387-92

67. Newbrun E, Pigman W (1960) The hardness of enamel and dentine. Aust Dent J 5:210-217

68. Niinomi M (2002) Recent metallic materials for biomedical applications. Metall Mater Trans A 33:477-486

69. Nixon GS, Livingston HD, Smith H (1966) Estimation of manganese in human enamel by activation analysis. Arch Oral Biol 11:247-252

70. Okazaki M, Takahashi J, Kimura H (1986) Crystallinity and solubility behavior of iron-containing fluoridated hydroxyapatites. J Biomed Mater Res 20:879-886

71. Oke O (1964) Chemical studies on corchorus. Indian J Med Res 52:1266

72. Panzera C, Kaiser LM (1999) Dental porcelain composition.US Patent 5,944,884, 31 Aug 1999.

73. Paul S, Peter A, Pietrobon N, Hämmerle CHF (2002) Visual and spectrophotometric shade analysis of human teeth. J Dent Res 81:578-582

74. Prasad M, Saxena S, Amritphale SS (2001) Adsorption models for sorption of lead and zinc on francolite mineral. Ind Eng Chem Res 41:105-111

75. Ramesh S, Tan CY, Peralta CL, Teng WD (2007) The effect of manganese oxide on the sinterability of hydroxyapatite. Sci Tech Adv Mater 8:257-261

76. Reitznerová E, Amarasiriwardena D, Kop áková M, Barnes RM (2000) Determination of some trace elements in human tooth enamel. Fresenius J Anal Chem 367:748-754

77. Ren F, Leng Y, Xin R, Ge X (2010) Synthesis, characterization and ab initio simulation of magnesium-substituted hydroxyapatite. Acta biomater 6:2787-96

78. Rey C, Collins B, Goehl T, Dickson I, Glimcher M (1989) The carbonate environment in bone mineral: a resolution-enhanced Fourier transform infrared spectroscopy study. Calcified tissue international 45:157-164

79. Ribeiro CC, Gibson I, Barbosa MA (2006) The uptake of titanium ions by hydroxyapatite particles-structural changes and possible mechanisms. Biomaterials 27:1749-61

80. Roman A, Chicot D, Lesage J (2002) Indentation tests to determine the fracture toughness of nickel phosphorus coatings. Surf Coat Technol 155:161-168

81. Schroeder HA, Vinton WH, Balassa JJ (1963) Effect of chromium, cadmium and other trace metals on the growth and survival of mice. J Nutr 80:39-47

82. Sekino M, Nakagawa H, Iwamoto O, Ushioda M (2001) Dental porcelain.patent 6,187,701, 13 Feb 2001.

83. Shukla GS, Singhal RL (1984) The present status of biological effects of toxic metals in the environment: lead, cadmium, and manganese. Can J Physiol Pharmacol 62:1015-1031

84. Sighinolfi GP, Gorgoni C, Bonori O, Cantoni E, Martelli M, Simonetti L (1989) Comprehensive determination of trace

elements in human saliva by ETA-AAS. Microchim Acta 97:171-179

85. Simmons LM, Al–Jawad M, Kilcoyne SH, Wood DJ (2011) Distribution of enamel crystallite orientation through an entire tooth crown studied using synchrotron X–ray diffraction. Eur J Oral Sci 119:19-24

86. Sprawson E, Bury FW (1928) On the chemical evidences of the organic content of human enamel. Proc R Soc Lon B Biol Sci 102:419-426

87. Tang XL, Xiao XF, Liu RF (2005) Structural characterization of silicon-substituted hydroxyapatite synthesized by a hydrothermal method. Mater Lett 59:3841-3846

88. Wang J, Nonami T, Yubata K (2008) Syntheses, structures and photophysical properties of iron containing hydroxyapatite prepared by a modified pseudo-body solution. J Mater Sci Mater Med 19:2663-2667

89. Weir A, Westerhoff P, Fabricius L, Hristovski K, von Goetz N (2012) Titanium dioxide nanoparticles in food and personal care products. Environ Sci Technol 46:2242-2250

90. White SN, Paine ML, Luo W, et al. (2000) The dentino-enamel junction is a broad transitional zone uniting dissimilar bioceramic composites. J Am Ceram Soc 83:238-40

91. Xue J, Zhang L, Zou L, et al. (2008) High-resolution X-ray microdiffraction analysis of natural teeth. J Synchrotron Radiat 15:235-238

92. Yamauchi M, Sakai M, Kawano J (1988) Clinical application of pure titanium for cast plate dentures. Dent Mater J 7:39

93. Zhang J, Wang D, Zhou J, Yao A, Huang W (2010) Precise adsorption behavior and mechanism of Ni(II) ions on nano-hydroxyapatite. Water Environ Res 82:2279-84

94. Zumdahl SS, Zumdahl SA (2000) Chemistry. Boston: Houghton Mifflin.

Groundwater Remediation Design Using Physics-Based Flow, Transport, and Optimization Technologies

Larry M Deschaine[1, 2], Theodore P Lillys[1], and János D Pintér[3]

[1]HydroGeoLogic, Inc., Reston, VA 20190, USA

[2]Department of Energy & Environment, Chalmers University of Technology, Göteborg, SE 412 96, Sweden

[3]PCS Inc, Halifax, NS, Canada

ABSTRACT

Background

The purpose of this work was to demonstrate an approach to groundwater remedial design that is automated, cost-effective, and broadly applicable to contaminated aquifers in different geologic settings. The approach integrates modeling and optimization for use as a decision support framework for the optimal design of groundwater remediation systems employing pump and treat and re-injection technologies. The technology resulting from the implementation of the methodology, which we call Physics-Based Management Optimization (PBMO), integrates physics-based groundwater flow and transport models, management science, and nonlinear optimization tools to provide stakeholders with practical, optimized well placement locations and flow rates for remediating contaminated groundwater at complex sites.

Results

The algorithm implementation, verification, and effectiveness testing was conducted using groundwater conditions at the Umatilla Chemical Depot in Umatilla, Oregon, as a case study. This site was the subject of a government-sponsored remedial optimization study. Our methodology identified the optimal solution 40 times faster than other methods, did not fail to perform when the physics-based models failed to converge, and did not require human intervention during the solution search, in contrast to the other methods. The integration of the PBMO and Lipschitz Global Optimization (LGO) methods with standalone physically based models provides an approach that is applicable to a wide range of hydrogeological flow and transport settings.

Conclusions

The global optimization based solutions obtained from this study were similar to those found by others, providing method verification. Automation of the optimal search strategy combined with the reliability to overcome inherent difficulties of non-convergence when using physics models in optimization promotes its usefulness. The application of our methodology to the Umatilla case study site represents a rigorous testing of our optimization methodology for handling groundwater remediation problems.

BACKGROUND

The increasing scarcity and degradation of potable water resources is an issue of global concern. Sources of water quality contamination include releases of chemical and radionuclide contaminants from point-source and non-point source origins. The costs of addressing water quality issues on a global scale are substantial. Regulatory agencies such as the U.S. Environmental Protection Agency (EPA) and counterpart agencies in other countries must consider economic and human health costs with budgetary constraints in their efforts to clean up contaminated groundwater and restore water resources to beneficial reuse. The EPA documents this need to perform groundwater remediation in an optimal manner in the *"National Strategy to Expand Superfund Optimization Practices from Site Assessment to Site Completion"* (USEPA, 2012).

The objective of this research is to develop and demonstrate an automated, cost-effective, and broadly applicable approach to groundwater remedial design applicable to contaminated aquifers in different geologic settings. This paper presents the development, scope and application of the simulation-optimization approach for remediating contaminated groundwater including reliability and efficiency verification to find globally optimized solutions to groundwater remediation problems. The approach is demonstrated using a well-studied, publically documented site example that was the subject of a government-sponsored remedial optimization study.

During the past two decades, researchers and engineers have been seeking methods for optimizing the design of ground water treatment systems. Table 1 provides an overview of existing algorithms and methods along with their salient features. Recognizing the limitations of the previously used optimization approaches and the need for efficient and effective groundwater remediation optimization tools motivates this research. This new approach combines flexible and efficient optimization techniques with commonly used subsurface groundwater flow and transport simulation models.

Table 1: Key features and limitations of previous optimization tools

Tool identification	Key features and limitations
GWM: Ground-Water Management Process for the U.S. Geological Survey (USGS) MODFLOW-2000 (Ahlfeld, et al., 2005)	Performs optimization using Linear Programming (LP) or Sequential Linear Programming (SLP).
	Tightly integrated to the MODFLOW code.
	Handles only confined flow and mildly non-linear unconfined flow situations.

MGO: Modular Groundwater Optimizer (Zheng and Wang, 2003) based on MODFLOW and the MT3DMS code (Zheng and Wang, 1999) for contaminant transport simulation	Performs optimization using heuristic global optimization methods, including Genetic Algorithm (GA) and Tabu Search (TS).
	Tightly integrated to the MODFLOW and MT3DMS codes.
	Computationally burdensome and cumbersome to use even for relatively straightforward practical situations.
SOMOS: Simulation/Optimization Modeling System (Peralta, 2004)	Performs optimization using a combination of GA, TS, and Artificial Neuron Network (ANN) in conjunction with groundwater flow and solute transport modeling.
SEA: Successive Equimarginal Approach, a hybrid of the gradient-based method and the deterministic heuristic-based method (Guo, et al., 2007)	Performs optimization using SEA to alleviate some of the computational burden of MGO.
	Integrated with MODFLOW and MT3DMS.
	Cumbersome to use requires frequent user intervention and may not lead to a global optimum.

Deschaine et al.

Deschaine et al. Environmental Systems Research 2013 2:6, doi:10.1186/2193-2697-2-6

DoD/ESTCP Simulation-Optimization Demonstration Project

This approach is tested using a study problem posed as part of the joint U.S. Department of Defense (DoD)/Environmental Security Technology Certification Program (ESTCP) Groundwater Remediation Optimization Study (Minsker et al. 2004); the groundwater contamination remediation design at the Umatilla Chemical Depot in Umatilla, Oregon. The site has a pre-existing and operational remedy-in-place (RIP) installed to remediate the Royal Demolition Explosive (RDX) and 2,4,6-Trinitrotoluene (TNT) contaminated groundwater plumes. The RIP consists of a groundwater pump and treat remediation system.

The original research teams used three optimization approaches during the DoD/ESTCP study. The DoD/ESTCP report presents these approaches in their entirety in Appendix D, Volume II. The design approaches consisted of Subject Matter Expertise (SME); the Modular Groundwater Optimization (MGO) (Zheng and Wang 2002), and; the Simulation/Optimization Modeling Optimization Software System

(SOMOS) (Systems Simulation/Optimization Laboratory SSOL2002). The team from the University of Alabama applied the MGO approach, the team from Utah State University applied the SOMOS approach, and the group from GeoTrans applied an SME-based subjective engineering approach. The publically available project web site (http://www.frtr. gov/estcp) provides the DoD/ESTCP study reports, groundwater flow and transport models, and modeling files of the final solutions to this problem.

The Umatilla research problem formulation used for testing is ESTCP Formulation 1. The goal of the formulation aims to reduce the projected clean up times of RDX and TNT at minimal cost, subject to constraints on the total allowable pumping and injection, treatment capacity, and the number of new wells needed. The experimental design of the DoD/ESTCP study directed the investigators to consider the existing flow and transport models as "up-to-date and acceptable for design purposes". The groundwater flow model code is USGS MODFLOW-96 (McDonald and Harbaugh1988; Harbaugh and McDonald 1996). The model code MT3DMS4 (Zheng and Wang 1999) for multispecies contaminant transport. The objective function calculator provided by ESTCP evaluates the cost associated with a remedial design and its performance. U.S. Army Corps of Engineers USACE (1996) developed and provided the MODFLOW-96 groundwater flow and MT3DMS4 solute transport models to the ESTCP study group.

The algorithm developed in this study - Physics-Based Management Optimization (PBMO) – is an automated simulation-optimization based method. Examination of the optimal design solution from PBMO with those from the ESTCP study demonstrates its effectiveness. The PBMO solution acceptance metric is a cost equal to or lower than developed by the MGO team. MGO provided one of the lowest costs with the least amount of computational effort of the automated approaches. Using the same physically based models and objective function calculator in all the studies isolates the performance of the demonstrated optimization algorithms.

Umatilla Chemical Depot, Umatilla Oregon Site Background and Description

Briefly, Umatilla is a 19,728-acre military reservation established in 1941 as an ordnance depot for the storage, renovation, and demilitarizing of conventional munitions, and for the storage of chemical munitions. As of 1994, the Umatilla site only stored chemical munitions awaiting destruction. A washout plant operated at the site in the 1950s and 1960s. Discharges to unlined lagoons consisted of an estimated 85 million gallons washout water laden with RDX and TNT. The water table is present about 47 feet beneath the bottoms of the lagoons. The resulting soil and groundwater contamination caused the placement of Umatilla on the EPA National Priorities List (NPL) in 1984. Section 3.1.1 of the study report (Minsker et al. 2004) provides additional description of the Umatilla site and historical summary.

The Record of Decision (U.S. Army Corps of Engineers USACE 1994) – the legal document governing the remediation approach - specified a pump and treat system with reinjection of treated groundwater as the remedial alternative. The treatment system commenced operations in January 1997; this action is the RIP. The system comprised three active extraction wells [EW-1, EW-3 and EW-4], three active infiltration/recharge basins [IF-1, IF-2 and IF-3], and a granular activated carbon (GAC) treatment system with a capacity of 1,300 gallons per minute (gpm). Figure 1 provides the locations of the system components. The extraction well (EW-2) installed 100 feet northwest of EW-4 is not part of the RIP, but is available for inclusion as merited.

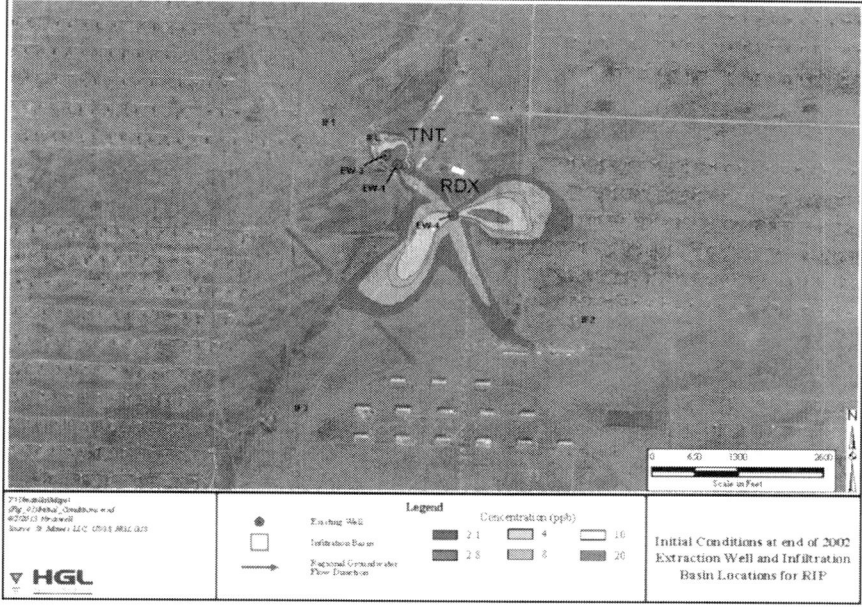

Figure 1: Initials conditions at end of 2002 extraction well and infiltration.

The initial RIP included an existing industrial lagoon designated IFL. However, suspicions that infiltrating water could spread the TNT plume rather than flushing the unsaturated zone of contamination resulted in discontinuing its use. By mid-July 1999, the RIP system had extracted, treated and recharged approximately 1.27 billion gallons of groundwater and removed an estimated 3,000 kilograms (kg) (6,614 pounds) of RDX and 400 kg (882 pounds) of TNT. The simulated RDX and TNT contaminant plumes, shown in Figure 2, represent the extent of contamination in the shallow aquifer at the end of 2002. The maximum RDX and TNT concentrations at that time were 28.2 and 86.7 micrograms per liter (µg/L), respectively.

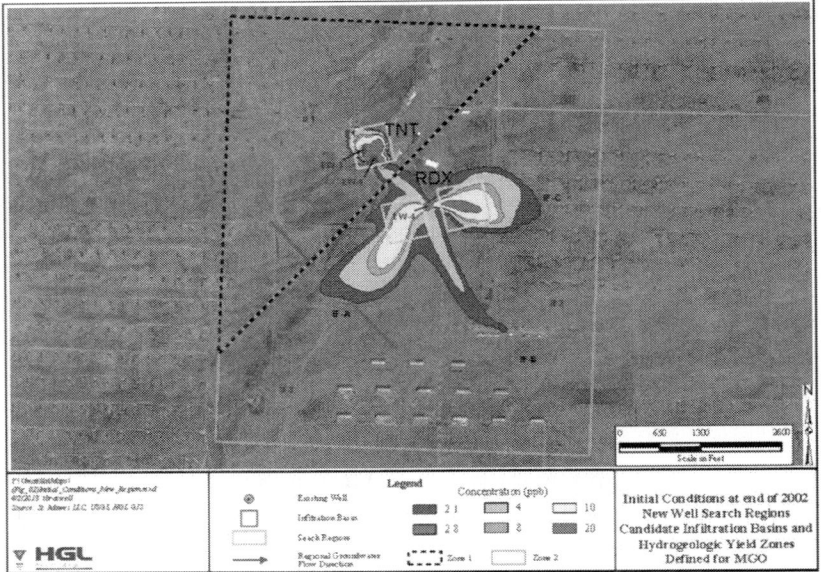

Figure 2: Initials conditions at end of 2002 new well search regions candidate infliction basins and hydrogeologic yield zones defined for MGO.

METHODS

Strategic planning for groundwater remediation, water resources planning and dewatering for mineral and resource mining requires tools that incorporate the complex constraints associated with the environment and support decision making with varying levels of uncertainty. This can include uncertainty in the conceptual site model, subsurface characterization (e.g. geologic material location and properties), chemical transport (e.g. reactions, natural attenuation, biodegradation), and funding levels (e.g. annual and total life cycle project) (ITRC, 2007).

Physics-based models: MODFLOW, MT3D, MODFLOW-SURFACT, and; MODHMS (HydroGeoLogic, Inc. HGL 2012) are highly effective at integrating remedial technology selection and application with competing remediation goals among regulators, site owners/custodians, and other stakeholders. The reasons for this are:

- *Comprehensive*: Physics-based models incorporate realistic and measurable physical parameters required for accurately describing the fate and transport of dissolved contaminants within aquifers. Optimal solutions based on physics-based models are more reliable than those from "lumped" parameter or ad hoc approximations.

- *Efficient and Effective*: Physics-based models are superior to lumped parameter or ad hoc models in accounting for changes in contaminant mass within aquifers. Increased accuracy yields solutions that better manage the use of remedial technologies and optimize costs and, therefore, be in complete control of the utilization of available funding.

- *Flexible*: Physics-based models can readily incorporate extended physical analysis to enable the development of optimal planning scenarios for processes that are outside of historical (data) observations, whereas models built on regression, interpolation, or extrapolation methods may not be representative.

Groundwater remediation problems amenable to physics-based decision optimization include the development of cost-effective and sustainable remedial designs; RIP evaluation and operation and maintenance (O&M) optimization; the development of optimized exit strategies to minimize life-cycle costs; the development of site completion/closure strategies; and water quality management issues for riparian and lacustrine settings, wetlands, and estuaries. This approach provides decision support to program and project managers regarding the best methods for remediating a site with contaminated groundwater.

Cost minimization is the overall optimization objective. In the case of groundwater remediation, well locations and their extraction or injection rates are the decision variables, and the simulated hydraulic heads and contaminant concentrations in groundwater are the state variables. *Decision variables* are the design elements of the problem studied. *State variables* represent the resultant simulated system. The *objective function* is a systematic accounting of costs which include the number, operation, and maintenance of the design elements (decision variables) and account for the changes in the modeled system (state variables) response from the candidate design. Computed over the life cycle of the simulation, the objective function represents the total remediation cost. The incorporation of *constraints* ensures the solution is practical and implementable. Constraints capture mechanical or physical process limitations (such as maximum pumping rates, and treatment train capacities) or interim and final values of state variables at discrete or continuous regions of the modeled domain. Assessment of *candidate solutions* is conducted by examining the cost value and the constraint compliance. A *single model function evaluation* is the cycle of the decision variable selection, process simulation, model objective function evaluation and constraint evaluation of a candidate design. *Optimization* is the automatic, guided process of the decision variable selection and application through repeated model function evaluations to arrive at the optimal objective function value that satisfies all constraints. The automatic adjustment of decisions variables terminates (ideally) at the global optimum (least cost).

PBMO's approach provides decision support for groundwater remediation, water resources planning and dewatering for mineral and resource mining via flexible integration of physics-based (groundwater flow and transport) models and optimization algorithms. Modular design promotes flexibility via independence of the physics model and optimization method. Linking the appropriate physics-based simulator with the best optimization algorithm(s) enables the solution to a wide range of problems types. Figure 3 shows how the modular design links the global optimization algorithms with the appropriate physics-based simulators to develop an optimal strategy. In this work, the first (top) half of the medallion represents the HGL_OPT optimization algorithms which include the Lipschitz Global Optimization (LGO©) solver suite (Pintér 1996 2002 2009 2013). The second (bottom) half represents the physically based numerical groundwater flow and transport simulators USGS MODFLOW-96 and MT3DMS4B (in this study); however any process simulator with a text interface is implementable.

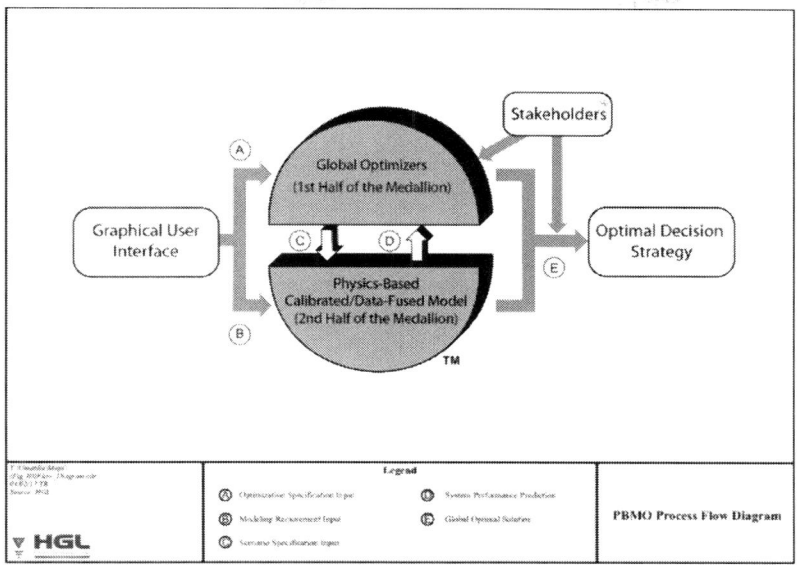

Figure 3: PBMO process flow diagram.

Subsurface simulators represent the physical processes of flow and transport in a mathematical form. Problem-specific, physics-based models represent the groundwater processes/conditions under investigation. From an optimization perspective, these simulation models can be *"black-boxes"*, that is models that receive input to produce an output without revealing the process. Since these groundwater flow and transport processes can occur in saturated or variably saturated conditions, in porous or fractured media, the numerical model used for the simulation will have a linearity *category*; the simulation can be linear, mildly non-linear (assumedly or provably convex), or highly non-linear (assumedly or provably non-convex). The system under consideration can be either single phase (water) or multiphase (water-NAPL-gas). The transport processes can be represented simply by using advective particle tracking, as a single non-reactive solute that undergoes dispersion and diffusion, or as a multicomponent system whose constituents interact with each other and the surrounding environment and are subject to hysteresis. The equations that govern these processes are Lipschitz-continuous, whose state-space representation can range from elliptic to parabolic or nearly hyperbolic when advection dominated. When no or low diffusion and dispersion are present relative to advection, steep fronts or shocks occur in the solutions. A complication for efficient optimization approach design arises when that the decision variables used in a candidate model evaluation during optimization change the model linearity category. Typically, the simulation of groundwater flow with slight drawdown is linear to mildly non-linear (because the saturated aquifer thickness does not change appreciably), whereas the simulation of significant groundwater drawdown or coupled groundwater flow and transport are most often highly non-linear. For the model scenario considered in this study, the flow system will range from linear far from the remediation system and contaminant plumes to highly non-linear in the contaminant plumes and near the remediation wells. The objective function is non-linear.

The USGS numerical simulator, MODFLOW-96, solves the following 3-D governing partial-differential equation for transient, saturated groundwater movement through porous media to yield a spatial distribution of the potentiometric head as a function of time:

$$\frac{\partial}{\partial x}\left(K_{xx}\frac{\partial h}{\partial x}\right) + \frac{\partial}{\partial y}\left(K_{yy}\frac{\partial h}{\partial y}\right) + \frac{\partial}{\partial z}\left(K_{zz}\frac{\partial h}{\partial z}\right) - q_s$$

$$= S_s\frac{\partial h}{\partial t} \tag{1}$$

Where K_{xx}, K_{yy} and K_{zz} are the values of hydraulic conductivity [L/T] along the x, y, and z coordinate axes, respectively, assumed to be parallel to the major axes of hydraulic conductivity;h is the potentiometric head [L]; q_s is a volumetric flux per unit volume of the aquifer and represents sources (injection wells) and/or sinks (extraction wells) of water [1/T]; S_s is the specific storage of the aquifer material [1/L]; and t is time (T).

Injection wells represent the infiltration/recharge basins. The initial ESTCP study used some terms interchangeably: injection, infiltration, and recharge. For clarity with the initial study language, we maintain that terminology in this paper to refer to the return of treated groundwater to the subsurface. The distribution of heads (one of the state variables), the source/sink data and, the candidate decision variables used or generated from the MODFLOW-96 simulation provide input to MT3DMS to solve the partial differential equation describing the fate and transport of contaminant species k in 3-D, transient groundwater flow systems. The governing equation is:

$$\frac{\partial\left(\theta C^k\right)}{\partial t} = \frac{\partial}{\partial x_i}\left(\theta D_{ij}\frac{\partial C^k}{\partial x_j}\right) - \frac{\partial}{\partial x_i}\left(\theta v_i C^k\right) + q_s C_s^k$$

$$+ \sum R_n$$

(2)

Where θ is the porosity of the aquifer material [-]; C^k is the dissolved concentration of species k [M/L^3]; $x_{i,j}$ is the distance along the respective coordinate axis, x, y or z [L]; D_{ij} is the hydrodynamic dispersion coefficient tensor [L^2/T]; vi is the seepage or linear pore water velocity [L/T]; C_s^k is the concentration of the source or sink flux for species k [M/L^3], and; $\sum R_n$ is the chemical reaction term [M/(L^{3}T)].

Darcy's Law couples the transport and flow equations.

$$v_i = -\frac{K_i}{\theta}\frac{\partial h}{\partial x_i}$$

(3)

McDonald and Harbaugh (1988) provide details on MODFLOW. Zheng and Wang (1999) provide details for MT3DMS.

To provide an optimization solver with the capability and efficiency to address this range of model types, PBMO is developed and implemented leveraging a suite of optimization algorithms that efficiently solve the various formulations expected to occur in the optimal decision support approaches presented here.

HGL_OPT is the master optimization driver. It comprises machine learning, objective function estimating methods, and SME-developed heuristics specifically useful for quickly developing high quality solutions to complex problems such as contaminant flow and transport remedial design (Deschaine 1992 and Deschaine 2003; Deschaine and Pintér 2003; and Deschaine et al. 1998 2001and Deschaine et al. 2011). These solutions can be used to focus (or initialize) the LGO-specific global and local optimization solvers.

The optimization algorithms included in the optimizer suite include linear programming (LP), sequential linear programming (SLP), sequential linear approximation (SLA), sequential quadratic programming (SQP), generalized reduced gradient (GRG), outer approximation (OA), Branch and Bound (BB), Globally Adaptive

Random Search (GARS), and Multi-start Random Search (MS). Specifically, LP solves linear models; OA, SLP, SLA, SQP, and GRG serve to deal with mildly non-linear models; BB, GARS, and MS—in proper combination with SLA, SQP, and GRG—serve to handle highly nonlinear models.

Currently, the optimization system can handle model formulations with up to a few hundred binary (yes/no) decision variables, several thousand continuous variables, and a several thousand general constraints (in addition to variable bound constraints). The actual configuration can be adjusted to user demands. For example, it is possible to optimize models with 200 decision variables and 5,000 general constraints, or vice versa. Even larger model sizes can also be handled if the computer random-access memory (RAM) and the compiler used support this. The BB solver of LGO employs a mild assumption of Lipschitz continuity, which allows an efficient and a robust search for a global optimal value through a systematic partitioning and exploration of the entire feasible region. The GARS and MS solvers assume basic model function continuity. The optimization suite solves the following general model:

- Minimize the [arbitrary linear or nonlinear] objective function $f(x)$

- Subject to bound constraints and [arbitrary linear or nonlinear] constraints:

$$x \in D := \left\{ x_l \leq x \leq x_u ; g(x) \leq 0 \right\} \tag{4}$$

where x is a decision vector, an element of the real n-space R^n; $f(x)$ is a continuous objective function, $f{:}R^nR$, $(R=R^1)$; D is a non-empty set of admissible decisions, a subset of R^n. As shown by (4), the set D is defined by l, u which is an explicit, finite n-vector bound of x (a "box") in R^n; and $g(x)$ which is an m-vector of additional continuous constraint functions, $g{:}R^nR^m$.

The assumption is that Lipschitz continuity holds as expressed by the inequality

$$\left| f_j(x_1) - f_j(x_2) \right| \leq L_j \| x_1 - x_2 \| \tag{5}$$

where $[L]$ is the Lipschitz constant. This equation means that the variability of the values computed by the objective function calculator is bounded with respect to the variability of the system input variables $[x]$ by the Lipschitz constant. This condition is an expected property of all physically based models.

Effective development and specification of a problem for groundwater remediation projects requires SME understanding of the physical processes and their constraints. The options under consideration include selecting which of the physics-based processes that need modeled (such as aquifer quality remediation via bioremediation or pump-and-treat) along with the constraints that the solution must satisfy. The formulation below is directly applicable and extensible to a wide range of groundwater dewatering and remediation problems. The basic elements of the mathematical formulation consist of an objective function and constraints.

Minimize:

$$f(x) = \sum_{i=1}^{N} (\delta_i, \alpha_i, q_i)$$

(6a)

subject to:

$$\sum_{i=1}^{N} q_i \geq Q^* \ \forall i \in I$$

(6b)

$$q_i \leq q_i^* \quad \forall i \in I$$

(6c)

$$c_j \leq c_j^* \quad \forall j \in J$$

(6d)

$$h_k \geq h_k^* \quad \forall k \in K$$

(6e)

$$t \leq T_{max}$$

(6f)

Here, the objective function $f(x)$ is a simplified version of life cycle cost for illustrative purposes. It consists of the flow rate from extraction wells $[q_i]$, a unit cost to process and treat the water $[a_i]$; and it assesses if an extraction well is pre-existing or not by examining for $\forall i$ locations and assigning a $(0,1)$ multiplier if a well installation required $(\delta_i=1)$ or is pre-existing $(\delta_i=0)$. The actual formulations of the objective function for optimal design problems often involve quite extensive and detailed cost information and functions such as used in this study. The

constraint $\sum_{i=1}^{N} q_i \geq Q^* \forall i \in I$ requires the installed system pumps at least some minimum volume of water Q. The $q_i \leq q_i^*$ constraint sets upper limits on a flow at each of the ith candidate well. The $c_j \leq c_j^*$ constraints ensure aquifer remediation to a certain acceptable residual level for all j locations. The $h_k \geq h_k^*$ constraints set minimum allowable water table elevations in the aquifer at all k locations. The $t \leq T_{max}$ constraint limits the allowable time for the activity to achieve the specified goals. These general constraints, like the cost function, can be modified and adapted as dictated by the needs of the project. For example, one can use this framework to incorporate constraints for differential land subsidence due to dewatering by using differencing constraints and adding constraints on the maximum slope of the water table or land surface. Similar approaches are effective for assessing depressurization and geotechnical stability.

The goal of constrained global optimization is to find at least one point x^* within the feasible region that satisfies $f(x^*) \leq f(x)$ for all x or to show that such a point does not exist. If no solution exists, then the decision makers realize it is an infeasible problem formulation. This enables the problem be reformulated. It is critical to determine if a feasible solution does not exist when solving a complex decision problem. This is a valuable capability of the physics-based optimization methodology. Many other optimization methods, specifically including heuristic techniques, cannot determine that a problem is infeasible: as a result, users can waste precious time and money searching for solutions with no hope of finding a feasible solution.

Umatilla Optimization Problem Formulation

The three optimization problem formulations considered in the DoD/ESTCP study were:

Formulation 1: Minimize the cost to remediate RDX and TNT in 20 years or less using the current treatment plant maximum operating flow rate of 1,300 (gpm) as an upper limit on the total groundwater extraction rate;

Formulation 2: Same as Formulation 1, except the maximum treatment flow rate increased to 1,950 (gpm);

Formulation 3: Minimize the aggregate remaining mass of RDX and TNT in 20 years using the current treatment plant flow rate.

The PBMO approach addresses all these types of optimization formulations. This exercise focuses on finding a solution to the problem as stated by Formulation 1. Appendix D, Volume II of the DoD/ESTCP report (Minsker et al. 2004) provides the mathematical formulation of the problem statement in detail.

The objective function for Formulation 1 is specified as follows:

$$MINIMIZE\left(C_{CW} + C_{CB} + F_{CL} + F_{CE} + V_{CE} \right.$$
$$\left. + V_{CG} + V_{CS}\right) \qquad (7)$$

where:

C_{CW}: Capital costs of new wells

$$C_{CW} = (25 \times I_{EW2})^d + \sum_{i=1}^{N_Y} (75 \times N_{W_i})^d \qquad (8)$$

N_y is the modeling year when cleanup is achieved [yr] as defined by $[C_{RDX}] \leq 2.1 \ \mu g/L$ and $[C_{TNT}] \leq 2.8 \ \mu g/L$ as measured by the nodal concentration value in the top layer.

N_{Wi} is the total number of new extraction wells (except well EW-2) installed in year i. New wells may only be installed in years corresponding to the beginning of a 5-year management period. Capital costs do not apply to pre-existing extraction wells.

I_{EW2} is a flag indicator; 1 when well EW-2 first comes into service, 0 otherwise.

75 is the cost of installing a new well [K$].

25 is the cost of putting existing well EW-2 into service [K$].

d indicates the application of the discount function to yield Net Present Value (N_{PV}) defined as

$$N_{PV} = \frac{C}{(1+r)^{y-1}}$$

(9)

c is the cost

r is the annual discount rate [1/yr]

y is the value of i in the summation

C_{CB}: Capital costs of new infiltration basins

$$C_{CB} = \sum_{i=1}^{ny} (25 \times N_{B_i})^d$$

(10)

N_{Bi} is the total number of new infiltration basins installed in year i. New recharge basins may only be installed in years corresponding to the beginning of a 5-year management period. The infiltration flux is evenly distributed throughout the basin.

25 is the cost of installing a new recharge basin independent of its location [K$/yr].

F_{CL}: Fixed cost of labor

$$F_{CL} = \sum_{i=1}^{N_Y} (237)^d$$

(11)

237 is the fixed annual O&M labor cost [K$/yr].

F_{CE}: Fixed cost of electricity (lighting, heating, and the like).

$$F_{CE} = \sum_{i=1}^{N_Y} (3.6)^d$$

(12)

3.6 is the fixed annual electric cost [K$/yr].

V_{CE}: Variable electric costs of operating wells (extraction and injection)

$$V_{CE} = \sum_{i=1}^{N_Y} \sum_{j=1}^{N_{wel_i}} \left(C_{W_{ij}} \times I_{W_{ij}} \right)^d$$

(13)

N_{weli} is the total number of extraction/injection wells active in year i.

C_{Wij} is the electrical cost for well j in year i. Costs differ for wells depending on the extraction rates of well j in year i, Q_{ij}:

$$C_{W_{ij}} = 0.01 \left(Q_{ij} \right) \quad \text{for } 0 \text{ gpm} < Q_{ij} \leq 400 \text{ gpm}$$

$$C_{W_{ij}} = 0.025 \left(Q_{ij} \right) - 6 \quad \text{for } 400 \text{ gpm} < Q_{ij} \leq 1000 \text{ gpm}$$

I_{Wij} is a flag indicator; 1 if well j is active in year i, 0 otherwise.

V_{CG}: Variable costs of changing GAC units

$$V_{CG} = \sum_{i=1}^{N_Y} \left[\gamma(\bar{c}_i) \times m_i \right]^d$$

(14)

$\bar{c}_i$ is the average influent concentration [µg/L] (RDX plus TNT) into the treatment plant from all of the extraction wells in the year i calculated as:

$$\bar{c}_i = \frac{\sum_{j=1}^{N_{wel_i}} Q_{ij} \bar{c}_{ij}}{\sum_{j=1}^{N_{wel_i}} Q_{ij}}$$

(15)

$\bar{c}_{ij}$ is the average influent concentration [µg/L] (RDX plus TNT) from well j in the year i

$\Upsilon(\bar{c}_i)$ is the cost of mass removed [K$/kg] as a function of average influent concentration into the treatment plant in year i, calculated as:

$$\gamma(\bar{c}_i) = \frac{-0.5(\bar{c}_i) + 225}{1000} \tag{16}$$

m_i is the mass of contaminant removed [kg] during year i calculated as:

$$m_i = \sum_{j=1}^{N_{wel_i}} Q_{ij}\bar{c}_{ij} \times \beta \tag{17}$$

β is a conversion factor to produce the result in units of [kg/yr].

V_{CS}: Variable cost of sampling.

$$V_{CS} = \sum_{i=1}^{N_Y} [150 \times (A_i/I_A)]^d \tag{18}$$

I_A is the initial plume area in layer 1 of the model based on the extent of RDX and TNT as measured in January 2003 where RDX and TNT exceeded their respective cleanup goals (2.1 and 2.8 µg/L, respectively) [m²]

150 is the annual sampling cost (as of January 2001) and considers both labor and analysis costs [K$/yr]

Ai is the modeled plume area in layer 1 in year i. This is evaluated at the beginning of a 5 year management period [m²]. A_i is defined as:

$$A_i = \sum_{j=1}^{N_{col}} \sum_{k=1}^{N_{row}} [\Delta x_j \Delta y_k \times I_{C_{jk}}] \tag{19}$$

N_{col} is the number of grid cell columns in the x direction
N_{row} is the number of grid cell rows in the y direction
Δx_j is the width of the jth grid cell column [m]
Δy_k is the width of the kth grid cell row [m]
IC_{jk} is a flag where:

C_{RDX}^{jk} is the concentration of RDX in the grid cell with indices j and k

C_{TNT}^{jk} is the concentration of TNT in the grid cell with indices j and k

The Formulation 1 constraints are:

- The modeling period consists of four 5-year management periods beginning with January 2003 (i or $year = 1$).
- Modifications to the system may only occur at the beginning of each management period.
- Remediation in the top layer of the model must be achieved within 20 years (e.g., RDX $\leq$ 2.1 µg/L and TNT $\leq$ 2.8 µg/L everywhere in top model layer).
- The total pumping rate, adjusted for the average amount of uptime, cannot exceed the treatment capacity of 1,300 (gpm) in any stress period. Evaluation of this constraint occurs at the beginning of each 5-year *management period*. It is computed as:

$$Q^* \Big/ {}_\alpha \leq 1300 \text{ gpm} \tag{20}$$

Here

α is a coefficient that accounts for the amount of average uptime ($\alpha = 0.9$)

Q^* is the total modeled flow rate during a 5-year management period.

- The hydrology dictates the upper sustainable flow limit on extraction wells. Extraction wells in Zone 1 may pump at a maximum rate of 400 (gpm), whereas extraction wells in Zone 2 may operate to a maximum of 1,000 (gpm). See Figure 2 for definitions of Zones 1 and 2:

$$\text{If } Zone(j,k) = 1,$$
$$\text{then } q_{jk}^{*} \Big/ {}_\alpha \leq 400 \text{ (gpm)}$$
$$\text{else } q_{jk}^{*} \Big/ {}_\alpha \leq 1,000 \text{ (gpm)} \tag{21}$$

where:

Zone (j, k) is a function of the jth grid cell column and k^{th} grid cell row that returns 1 if model grid (j, k) corresponds to Zone 1, and returns 2 if (j, k) corresponds to Zone 2

q^*_{jk} is the modeled extraction rate at model grid location (j, k).

- It is unallowable for the extent of groundwater contamination to increase beyond initial conditions at any time during the remediation.
- Total pumping and infiltration rates must be balanced at the beginning of every management period

$$\left| \sum_{j=1}^{N_{wel_i}} Q_{ij} - \sum_{k=1}^{N_{rech_i}} Q_{kj} \right| \leq 1 \text{ gpm}, \forall i \in \{1, 6, 11, 16\}$$

(22)

where:

N_{rech_i} is the number of injection wells operating in year i.

In summary, the optimization problem can be stated as follows: find the combination of simulated extraction and injection well locations and their operating rates that minimize the cost of reducing RDX and TNT concentrations within a 20 year time horizon while satisfying all the constraints on well, remedy operations, and plume behavior. The remedial system designs use the calendar year 2003 as the starting point.

The groundwater flow and transport models provided for the study simulate 20 years with four management periods of five years each. The extraction and injection flow rates can vary across the management periods, but not within a management period. The locations of the extraction and injection infrastructure can only vary between candidate solutions, not between management periods.

Optimization Solution Approach

The optimization problem is solved by defining the regions to search for the globally optimal settings of the decision variables, and prescribe how to conduct the search. The decision variable search regions mimicked, as closely as possible, the regions established for new well locations and infiltration basins by the MGO team. The MGO team confined their search for new extraction well locations to regions where the plume densities were highest as shown on Figure 2. Each region contains the areas of the highest concentration for the three principal lobes of the two contaminant plumes. In addition to these search areas for extraction wells, locations for three new infiltration basins, IF-A, IF-B, and IF-C in the original study at the extent of the RDX plume to the east, southeast, and southwest as alternatives to the four pre-existing recharge basins. In all, a total of 11 candidate regions exist for locating decision variables: four extraction areas and seven injection/recharge areas; the locations as defined by the MGO team. These 11 regions make up the infrastructure search areas of the remedy used to alter the distribution of RDX and TNT in groundwater. Active remediation solutions require non-zero total extraction and injection fluxes; hence, there must be at least one extraction location and one injection location active for all viable candidate solutions.

Within the extraction locations, a well(s) can be located anywhere in the search box (we restrict the well position to a model node). Table 2 provides the box location and number of candidate position locations for one or two wells in each singular box.

Table 2: Number of candidate extraction well locations in each of the three search boxes

Location	Coordinates	Size	Number of potential extraction well locations(1)
Box 1	(40,53) to (63,73)	24×21	127,260
Box 2	(82,63) to (91,83)	10×21	22,155
Box 3	(82,90) to (90, 96)	9×7	2,016

Includes single and double well location combinations per box.

Deschaine et al.

Deschaine et al. Environmental Systems Research 2013 2:6, doi:10.1186/2193-2697-2-6

Allowing an extraction well(s) to be located in any of the three search boxes results in 302,253 potential extraction well location combinations. The 128 different candidate location configurations for the infiltration basins results in a total of 38,688,384 candidate infrastructure system designs. Simultaneously, when determining the infrastructure configuration design, the water flow rates are optimized. This magnitude of options illustrates the difficulty of finding the optimal solution either by random searching or by using the SME subjective engineering judgment.

The approach used to solve the Umatilla problem consisted of the following generalized automated search strategy:

- Evaluate the cost of the RIP. The RIP consists of three extraction wells and three infiltration basins. Store these results as a current minimum.
- Begin evaluation of different combinations of extraction wells and infiltration basins.

    1. Set number of evaluation epochs equal to one.

    2. Initial extraction location: Initiate the solution using the study maximum number of new wells (two), located in Box 1 (which contains the RDX and TNT plumes).

    3. Initial injection location: IFL located in Box 1, the innermost infiltration basin.

    4. Initiate search strategy. Begin to cycle through and test the 4 pumping areas and 7 injection areas for quality of solution, use extraction rate total = system maximum capacity of 1170 (gpm) (which is 1,300 (gpm) * 90% uptime) as done by the MGO team.

    5. Set search cycle a minimum number of evaluations (12).

    6. Upon finding a cost lower than current minimum cost:

        a. Select the solutions with the two lowest costs.

        b. Explore these regions by conducting brief local random searches (LRS) on each solution to (statistically) determine the most promising configuration of wells and basins. Perform initial analysis using the physically-based model. The optimization search progresses and generates an increasing number of function evaluations. Machine learning is invoked to provide candidate solution objective function evaluation estimation which reduces computation burden when model evaluations models are extensively time consuming.

    7. Store these results (configuration, rates, cost).

8. Evaluate best current solution.

    a. If solution improves, replace previous remedial design configuration and objective function value.

    b. Else, continue.

9. Initiate a global optimization analysis using the LGO search options of GARS followed by GRG; initialize using currently best identified solution.

    a. Upon finding a cost lower than current minimum cost:

    b. Store these results.

    c. Else, continue.

10. If termination criteria met,

    a. Stop.

    b. Else, increment number of evaluation epochs by 1 until the user-specified maximum number reached. Store and print results.

    c. Go to step 2.

PBMO's partition and exploration approach enables implementation of the search strategy be conducted on multiple central processing units (CPUs). However, this test used a sequential implementation to mimic ESTCP test conditions. The search continues until the termination criterion satisfied. The termination criterion can either be the targeted optimal cost, the total number of flow and transport simulations, the number of simulations since the optimal value was last improved, or the total simulation (CPU) time consumed. In this case, the termination criterion was the optimal total remedial cost as published by the MGO team. The initial starting point for the total flow rate is the maximum treatment plant flow rate 1,170 (gpm) and mimics the MGO team. Initially, all extraction wells in Box 1 equally pumped; all extracted water injected into the single infiltration gallery IFL, and new candidate extraction wells placed randomly during the infrastructure search phases.

RESULTS AND DISCUSSION

The total net present value cost from PBMO ($1,664,085) significantly improved upon the costs committed for remediation at the site by implementing the RIP ($3,836,285) and the design achievable via SME ($2,230,905). The PBMO results mimicked MGO ($1,664,395) and SOMOS ($1,663,841). Table 3 presents the well location strategies and cost results for RIP, SME, MGO and PBMO (since MGO and SOMOS were similar).

Table 3: Optimal pumping strategies found using SME, MGO and PBMO for formulation 1 at Umatilla compared with existing RIP design

Name	Location (Layer, Row, Column)	Pumping/injection rate (gpm)				
		RIP	Formulation 1 solutions			
			SME design		MGO	PBMO
			Stress period 1	Stress period 2	Stress period 1	Stress period 1
EW-1	(1,60,65)	−128	−280	−350	−307.5	−292.5
EW-2	(1,83,84)					
EW-3	(1,53,59)	−105		−360	−219.5	−292.5
EW-4	(1,85,86)	−887	−660			
New-1 (T&E)	(1,48,57)			−100		
New-2 (T&E)	(1,49,58)		−230	−360		
New-3 (MGO)	(1,48,59)				−360	
New-4 (MGO)	(1,48,55)				−283	
New-5 (PBMO)	(1,48,57)					−292.5
New-6 (PBMO)	(1,52,61)					−292.5
IF-L	*					
IF-1	*	233	282	585		
IF-2	*	405	405		380	390
IF-3	*	483	482		790	780
IF-4	*			585		
IF-A	*					
IF-B	*					
IF-C	*					
Total cost in net present value ($)		$3,836,285	$2,230,905		$1,664,395	$1,644,085

*See Figure 2 for location.

(a negative flow rate indicates pumping; positive indicates injections).

Deschaine et al.

Deschaine et al. Environmental Systems Research 2013 2:6, doi:10.1186/2193-2697-2-6

The termination criterion in the test was the MGO cost value. PBMO found a lower cost that MGO in fewer than 120 model evaluations. Additional optimization analysis performed during reliability testing produced multiple solutions with lower costs - as low as $1,663,240 – and with different new well extraction locations and different rates in the extraction wells. These subsequent findings illustrate the multi-extremal structure of this design problem, thereby calling for global optimization based solution approaches.

Simulation model reliability over the range of feasible inputs is essential for determining the globally optimal variable values. We observed that 10.7% of the viable candidate designs simulated in the groundwater flow and transport models during the optimization search failed to converge. This primarily occurred when the extraction well flow rates were set near the high end of the acceptable range and which caused flow model cells to dewater. In instances where model cells become dewatered, the groundwater flow simulator could not converge to a solution without the application of a non-physical lower bound on the head in the dewatered cell[a]. Given that the two simulators execute sequentially during a function evaluation (groundwater flow followed by fate and transport), the failure of the flow simulator to provide a convergent solution prevents the transport simulator from predicting contaminant distributions resulting in a lost candidate design evaluation cycle.

[a]This approach was explicitly applied to one of the other study sites in the DoD/ESTCP study – see Section 3.2.3.4 in Volume I of Minsker et al. (2004).

PBMO addresses issues that arise in model simulations due to these harsh modeling scenarios imposed by formal optimization. PBMO examines the model simulation solution time and the flow and transport mass balance errors. Automated solver parameter adjustments can take place if the solution becomes unstable or inefficient. If a model nevertheless fails to converge after a user-specified maximum number of numerical solver parameter setting attempts, penalty functions divert the optimal search from exploring this solution region. Hence, while the algorithm handles the non-convergent model issue, should the simulations be noted to fail to converge either at an appreciable rate or in regions of the search space where the suspected location of the optimal value, a more robust model code should be used. This will alleviate the risk of not locating the true globally optimal solution. This study design necessarily used the modeling system used by the other teams in spite of the 10.7% model simulation failure rate.

Table 4 presents a breakdown of the capital, and operations and maintenance costs for each of the four strategies. Figure 4 shows the well locations for RIP and T&E conducted by the SME's, and the optimal well locations determined by MGO and PBMO The PBMO solution places two new wells into the TNT plume, as did MGO.

Table 4: Breakdown of the capital and O&M costs of RIP, SME, MGO and PBMO designs

Cost components	RIP existing design	SME strategy	MGO optimal strategy	PBMO optimal strategy
Capital costs of new wells (CCW)	$ -	$ 133,764	$ 150,000	$ 150,000
Capital costs of new infiltration basins (CCB)	$ -	$ 19,588	$ -	$ -
Fixed costs of labor (FCL)	$ 2,805,552	$ 1,263,086	$ 882,410	$ 882,410
Fixed costs of electricity (FCE)	$ 42,616	$ 19,186	$ 13,404	$ 13,404
Variable costs of electricity for operating wells (VCE)	$ 251,405	$ 91,952	$ 48,394	$ 48,402
Variable costs of changing GAC units (VCG)	$ 16,338	$ 14,301	$ 11,700	$ 11,382
Variable costs of sampling (VCS)	$ 720,374	$ 689,028	$ 558,487	$ 558,487
Objective function value	$ 3,836,285	$ 2,230,905	$ 1,664,395	$1,644,085

Deschaine *et al.*

Deschaine *et al.* *Environmental Systems Research* 2013 2:6, doi:10.1186/2193-2697-2-6

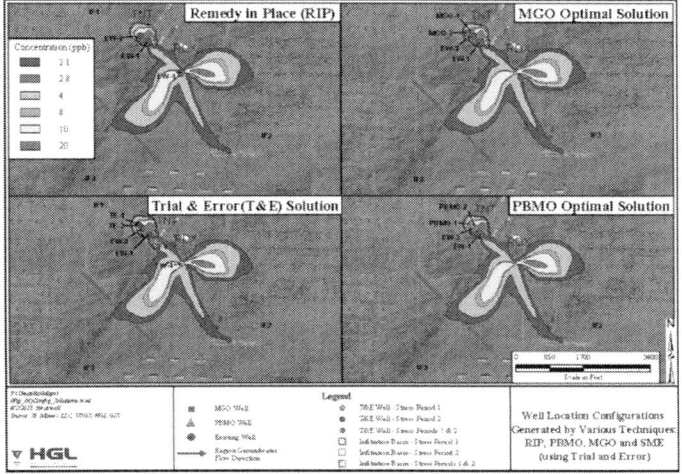

Figure 4: Well location configurations generated by various techniques: RIP, PBMO, MGO and SME (using trial and error).

Examining the performance comparisons of the algorithms[b] reported in the ESTCP study for the number of flow and transport simulations, PBMO found its solution using under 120 flow and transport simulations as compared to the estimated 5,000 simulations reported by the MGO investigators [see Volume II of the DoD/ESTCP report (Minsker et al., 2004)]. This is a significant improvement in efficiency, by an estimated factor of over 40. Furthermore, PBMO ran unattended, whilst MGO required numerous human interventions.

[b]The model files representing the optimal solution identified by MGO and SOMOS are on the web site, however, the input files to MGO or SOMOS to recreate the optimization study and generate the optimal solution are not available. Therefore, it was not possible to independently verify the reported operational performance of MGO or SOMOS. Hence, we use the reported values for the number of model runs, with MGO being the lesser number of ~5,000. To put this efficiency result in perspective, it would take nearly 5 days of clock time on a modern computer (~1.4 minutes per flow and transport simulation) to solve the problem using MGO, whereas PBMO found a somewhat

better solution in less than 3 hours. In addition, the MGO solution required several stops and starts as well as other interceding actions by the investigators to reach the final answer presented in the study. PBMO required a single execution without any human intervention: the optimization process was completely automated.

The authors believe that the observed increased efficiently lies in the integrated optimization search strategy. The PBMO approach supports rapid determination of "good" solution spaces that work synergistically with LGO, the core global optimization algorithm. The investigation of (Rios and Sahinidis 2012) supports this position via the results of testing 23 optimization algorithms, including LGO. The results of that comparative study indicate LGO to be orders of magnitude more efficient and reliable than the techniques selected by the original ESTCP study teams (see Figure 5, adapted from that study). Regarding the final solution to the Umatilla problem, the solutions generated by the optimization techniques are conceptually similar: all three utilize the same existing extraction wells, EW-1 and EW-3; both use the same infiltration basins, IF-2 and IF-3; and both locate new extraction wells near to the center of mass of the TNT plume, and remediate the site in 4 years. Figures 6, 7, 8, 9 show the RDX and TNT plumes at the end of Year 1 through Year 4— the time of remediation completion. Figure 10 and Figure 11 illustrate the maximum concentration of both contaminants over time in the top layer of the model. The (slight) over design by MGO compared with PBMO is evidenced regarding the TNT remediation results. The TNT remediation did not occur at the same time as the RDX remediation and the water quality was remediated cleaner than required by the project specifications.

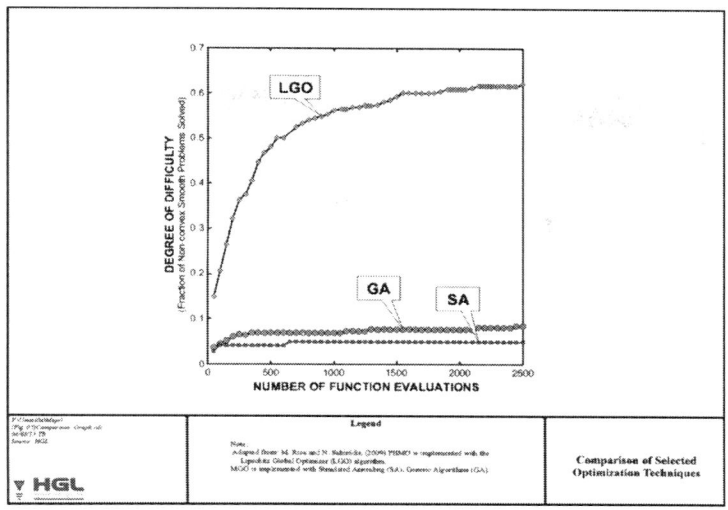

Figure 5: Comparison of selected optimization techniques.

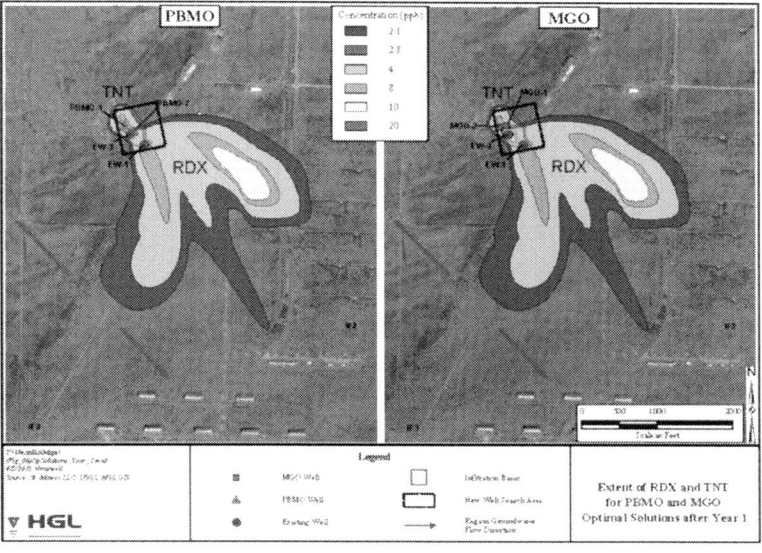

Figure 6: Extent of RDX and TNT for PBMO and MGO optimal solutions after year 1.

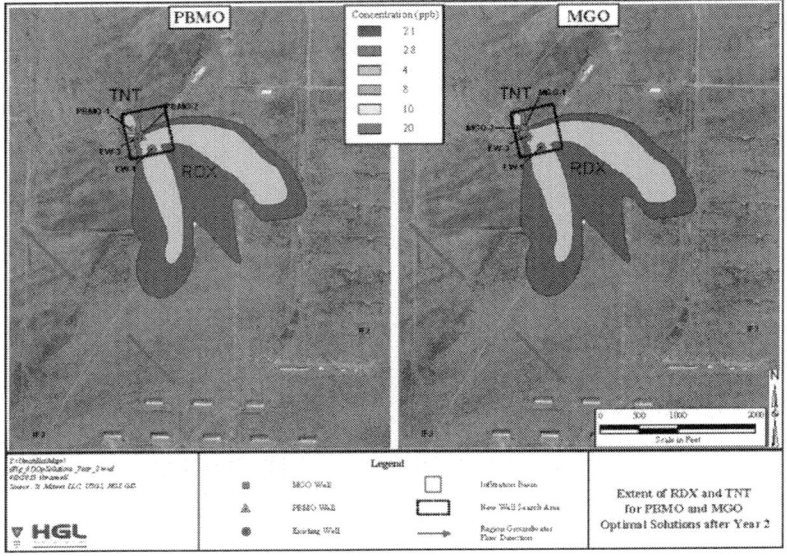

Figure 7: Extent of RDX and MGO optimal solutions after year 2.

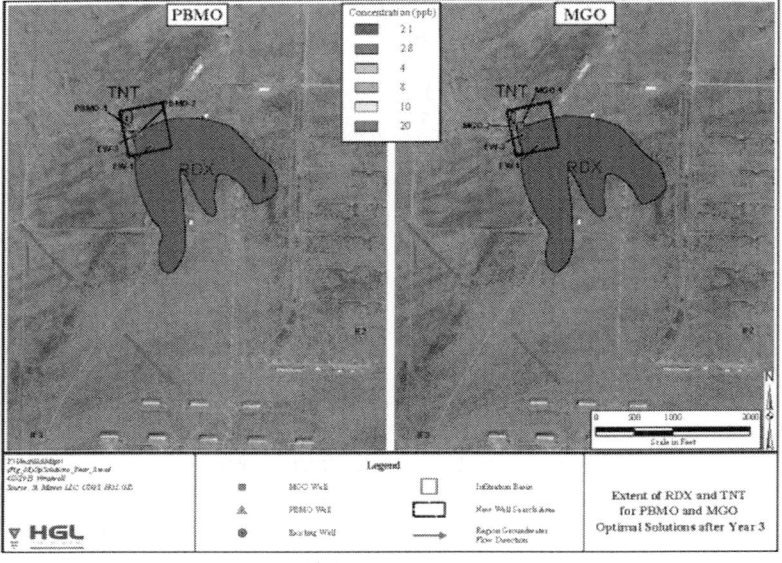

Figure 8: Extent of RDX and MGO optimal solutions after year 3.

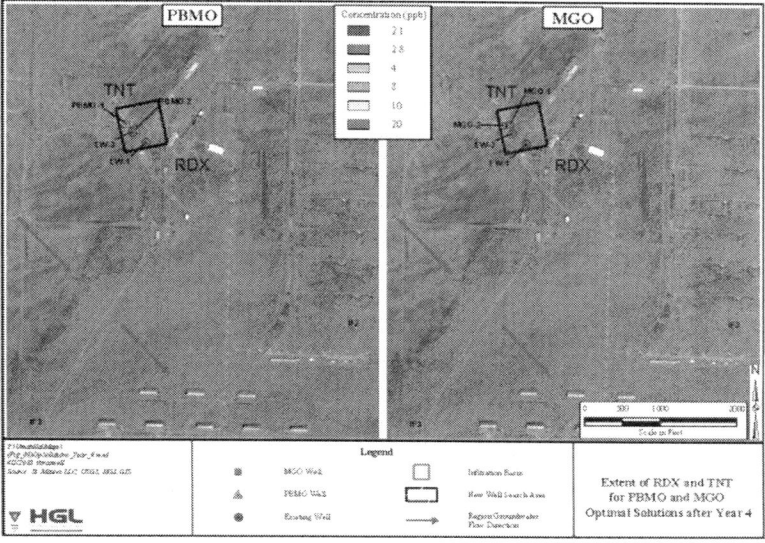

Figure 9: Extent of RDX and MGO optimal solutions after year 4.

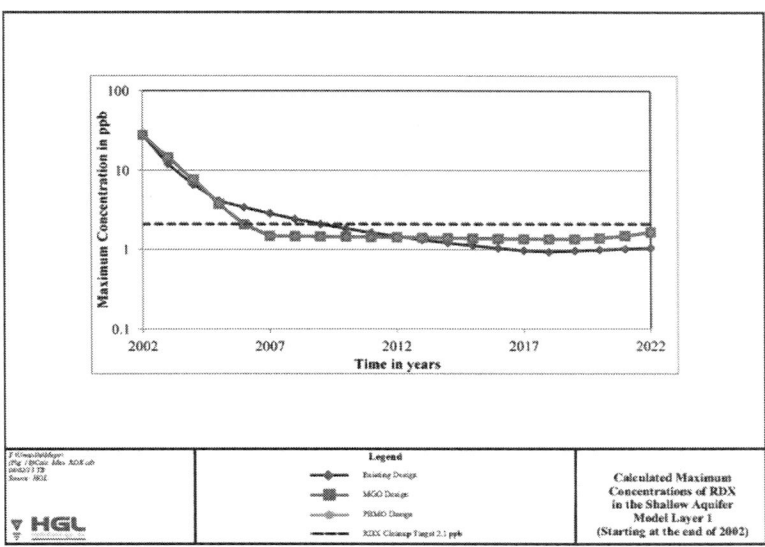

Figure 10: Calculated maximum concentrations of RDX in the shallow aquifer model layer 1 (Starting at the end of 2002).

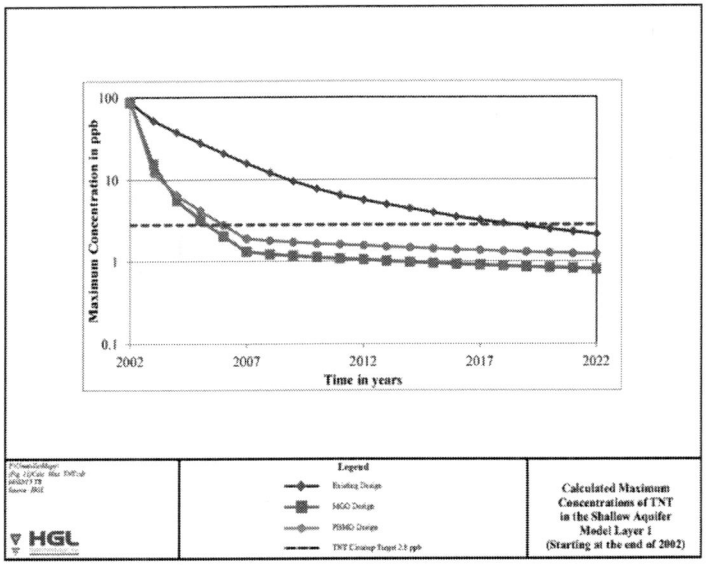

Figure 11: Calculated maximum concentrations of TNT in the shallow aquifer model layer 1 (Starting at the end of 2002).

Remediation Well Starting Position: Sensitivity Tests

The algorithm's reliability test consisted of assessing its ability to converge on the optimal result from different ill-placed starting locations. The well location search region used was the same search region defined for the northernmost box, the one that contains the initial RDX and TNT plumes as shown in Figure 2. However, instead of using the best configuration of the extraction well locations determined thus far (i.e. from the infrastructure search), the search box corners specify the initial new extraction well locations. The test scenarios consisted of six combinations of initial locations for the two new extraction wells at the four corners. Figure 12 shows the six different initial starting configurations for the two new extraction wells represented as the green triangles on the various corners. The results of interest are the final solution, the optimal placement of two new extraction wells to accompany EW-1 and EW-3, the number of simulation/optimization iterations required, and the elapsed time needed to achieve the solution. Table 5 presents the test results. Each of the six test cases found the same optimal locations. The number of model simulations varies between 124 and 127. In every scenario, identification of the optimal solution occurred in less than 3 hours of CPU time. In nearly 100 iterations, the GARS optimization algorithm found the optimal solution regardless that the search initiates from locations outside the contaminant plume. Reliability and minimal dependence on the starting solution initialization are fundamental and desirable features of a high-quality global solver.

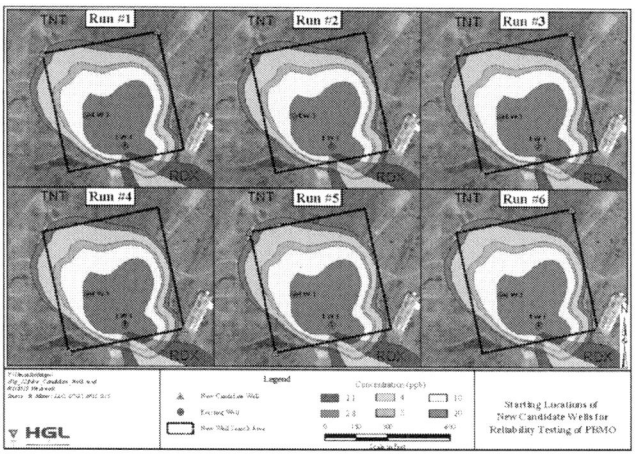

Figure 12: Starting locations of new candidate wells for reliability testing of PBMO.

Table 5: PBMO computational performance during robustness testing

Run #	Starting locations for new wells (L,R,C)	Optimal locations (L,R,C)	Optimal cost ($K)	# Flow/ transport iterations to optimal3	CPU time to optimal (min)1,2
1	(1,40,53)	(1,48,57)	1,664.1	124	164.3
	(1,40,73)	(1,52,61)			
2	(1,40,53)	(1,48,57)	1,664.1	124	164.5
	(1,63,73)	(1,52,61)			

3	(1,63,53)	(1,48,57)	1,664.1	127	177.8
	(1,40,73)	(1,52,61)			
4	(1,63,53)	(1,48,57)	1,664.1	127	178.3
	(1,63,73)	(1,52,61)			
5	(1,40,53)	(1,48,57)	1,664.1	124	161.7
	(1,63,53)	(1,52,61)			
6	(1,40,73)	(1,48,57)	1,664.1	124	161.0

[1]Intel Core i5 CPU 760 @ 2.80 GHz, 8 GB RAM, Win7 Pro.

[2]Intel Core i7 CPU 870 @ 2.93 GHz, 8 GB RAM, Win7 Pro.

[3]The number of iterations for each optimization run includes 1 evaluation for RIP and 24 evaluations for the optimal infrastructure configuration search.

Deschaine et al.

Deschaine et al. Environmental Systems Research 2013 2:6, doi:10.1186/2193-2697-2-6

CONCLUSIONS

The study demonstrates the effectiveness of an automated, cost-effective, and broadly applicable approach to groundwater remedial design. The application of the PBMO methodology to the Umatilla case study site represents a rigorous testing and validation exercise. This numerical analysis efficiently and automatically identified the best solution found by others. Extension of the approach for evaluation and optimal aquifer remediation management for different sites in different geologic settings is accomplished by using a site specific calibrated groundwater flow and transport model. The approach identified the optimal solution about 40 times faster than any of the other methods by reducing the number of time consuming flow and transport model evaluations. Comprehensive automation promotes efficiency, effectiveness and usability. Merit for reliably optimizing complicated systems is achieved through the ability to overcome the inherent non-

AUTHORS' CONTRIBUTIONS

The work presented here was carried out in collaboration between all authors. LD defined the research theme, designed and implemented the overall optimization approach. JP provided LGO optimization component and TL coded the objective function calculator. All authors have contributed to the paper preparation, and have seen and approved the manuscript.

ACKNOWLEDGEMENTS

This research was conducted in concert with LD's PhD studies in Complex Systems at Chalmers University, with financial support by HydroGeoLogic, Inc, 11107 Sunset Hills Road, Suite 400, Reston, VA 20190 (http://www.hgl.com). An animation of this case study has been prepared; see (http://www.hglsoftware.com/cleanup.cfm).

REFERENCES

1. Ahlfeld DP, Barlow PM, Mulligan AE (2005) GWM – a ground-water management process for the US Geological Survey modular ground-water model (MODFLOW – 2000). US Geological Survey. Open-File Report 2005-1072, 124 Pages, 2005.https://water.usgs.gov/nrp/gwsoftware/mf2005_gwm/OFR2005_1072.pdf

2. Deschaine LM (1992) Cost evaluation and optimization of ground water pump and treat programs, MSCE thesis. Storrs, Connecticut: University of Connecticut.

3. Deschaine LM, Ahlfeld DP, Ades MJ, O'Brien D (1998) An optimization algorithm to minimize the life cycle cost of implementing an aquifer remediation project - theory and case history. Society for Modeling and Simulation International, Simulators International XV, Simulation Series 30(3):53-58

4. Deschaine LM, Regmi S, Patel JJ, Fox TA, Ades MJ, Katyal A (2001) Design optimization of groundwater quality management challenges using the outer approximation method, The Society

for Modeling and Simulation International. Seattle, WA, USA: Advanced Simulation Technology Conference. pp pages 88-93 April 2001

5.   Deschaine LM (2003) Simulation and optimization of large-scale subsurface environmental impacts; investigations, remedial design and long term monitoring. Journal of Mathematical Machines and Systems, Kiev 3(4):Pages 201-218

6.   Deschaine LM, Pintér JD (2003) A comparison of the outer approximation method and lipschitz global optimization on optimal groundwater quality management. Atlanta, Georgia: Presented at the INFORMS Annual Meeting Conference. October 19-22, 2003

7.   Deschaine LM, Nordin JP, Pintér JD (2011) A computational geometric / information theoretic method to invert physics-based mec-model attributes for mec discrimination. Journal of Mathematical Machines and Systems, National Academy of Sciences of Ukraine, Kiev 2:50-61

8.   Guo X, Zhang CM, Borthwick JC (2007) Technical Report. Water Resources Research Journal 43(No. 8):WO8416-14 Pages August. http://onlinelibrary.wiley.com/doi/10.1029/2006WR004947/abstract

9.   Harbaugh AW, McDonald MG (1996) User's documentation for modflow-96, an update to the u.s. geological survey modular finite-difference ground-water flow model. Reston VA: U.S. Geological Survey Open-File Report 96–485. pp 56-485

10.   HydroGeoLogic, Inc. (HGL) (2012) MODFLOW-SURFACT™ Version 3.0; User's manual and guide. VA, USA. 20190

11.   ITRC (2007) In-Situ Bioremediation of Chlorinated Ethene DNAPL Source Zones: Case Studies. Prepared by The Interstate Technology and Regulatory Council, Bioremediation of Dense Non-Aqueous Phase Liquids (Bio DNAPL) Team. Refer to Chapter 9 "Simulation and Optimization of Subsurface Environmental Impacts; Investigations, Remedial Design and Long Term Monitoring of BioNAPL Remediation Systems".. 128-147 April 2007. http://www.itrcweb.org/Guidance/GetDocument?documentID=11

12.   McDonald MG, Harbaugh AW (1988) A modular three-dimensional finite-difference groundwater flow model, Techniques of Water Resources Investigations Book 6. Washington, D.C: U.S.

Geological Survey.

13. Minsker B, Zhang Y, Greenwald R, Peralta R, Zheng C, Harre K, Becker D, Yeh L, Yager K (2004) Final technical report for application of flow and transport optimization codes to ground water pumping and treat systems, Technical Report to the Environmental Security Technology Certification Program. Volumes I-III. TR-2237-ENV. California: Engineering Service Center, Port Hueneme. http://www.frtr.gov/estcp/estcp.htm

14. Peralta RC (2004) "SOMOS: Simulation/Optimization Modeling System". User's Manual, Software Engineering Division, Department of Biological and Irrigation Engineering. Logan, UT: Utah State University. p 48 Pages April 2004. http://www.frtr.gov/estcp/source_codes.htm

15. Pintér JD (1996) Global optimization in action. Kluwer Academic Publishers, Dordrecht Boston London, 1996. New York: Now distributed by Springer Science + Business Media.

16. Pintér JD (2002) Global optimization: software, test problems, and applications. In: Pardalos PM, Romeijn HE (eds) Handbook of Global Optimization, Volume 2, Dordrecht: Kluwer Academic Publishers. pp 515-569

17. Pintér JD (2009)Pardalos PM, Coleman TF (eds) Software development for global optimization, Providence, RI: American Mathematical Society.

18. Pintér JD (2013) LGO - A model development and solver system for global-local nonlinear optimization user's guide. Canada: Published and distributed by Pintér Consulting Services, Inc. http://www.pinterconsulting.com . First edition: June 1995; Current edition: March 2013

19. Rios M, Sahinidis N (2012) "Derivative-free optimization: A review and comparison of software implementations". J Glob Optim. http://link.springer.com/content/pdf/10.1007%2Fs10898-012-9951-y.pdf

20. Systems Simulation/Optimization Laboratory (SSOL) (2002) Simulation/Optimization Modeling System (SOMOS) Users Manual. Logan, UT: SS/OL, Biological & Irrig. Eng. USU. p 457

21. U.S. Army Corps of Engineers (USACE) (1994) Defense Environmental Restoration Program, Final Record of Decision,

Umatilla Depot Activity Explosives Washout Lagoons Ground Water Operable Unit. June 7, 1994

22. U.S. Army Corps of Engineers (USACE) (1996) Final Remedial Design Submittal, Contaminated Groundwater Remediation, Explosives Washout Lagoons, Umatilla Depot Activity, Hermiston Oregon. January 1996

23. USEPA (2012) National strategy to expand superfund optimization practices from site assessment to site completion. (September 2012) (PDF) OSWER 9200.3-75http://www.epa.gov/oerrpage/superfund/cleanup/postconstruction/optimize.htm

24. Zheng C, Wang PP (1999) MT3DMS: A modular three-dimensional multispecies transport model for simulation of advection, dispersion and chemical reactions of contaminants in groundwater systems; documentation and user's guide, Contract Report SERDP-99-1. Vicksburg, MS: U.S. Army Engineer Research and Development Center. available at http://hydro.geo.ua.edu/mt3d

25. Zheng C, Wang PP (2002) MGO – A modular groundwater optimizer incorporating modflow/mt3dms, documentation and user's guide. Draft. April 2002. http://www.frtr.gov/estcp/source_codes.htm

26. Zheng C, Wang P (2003) "Application of flow and transport optimization codes to groundwater pump-and-treat systems: Umatilla Army Depot, OR". Technical Report, Revised version 2/2003. Tuscaloosa, AL: University of Alabama. p pp. 41 http://www.frtr.gov/estcp/demonstration_sites.htm

The Bioliq®Bioslurry Gasification Process for the Production of Biosynfuels, Organic Chemicals, and Energy

Nicolaus Dahmen, Edmund Henrich, Eckhard Dinjus, and Friedhelm Weirich

Institute of Catalysis Research and Technology, Karlsruhe Institute of Technology (KIT), Campus Nord, Eggenstein-Leopoldshafen, D-76344, Germany

ABSTRACT

Background

Biofuels may play a significant role in regard to carbon emission reduction in the transportation sector. Therefore, a thermochemical process for biomass conversion into synthetic chemicals and fuels is being developed at the Karlsruhe Institute of Technology (KIT) by producing process energy to achieve a desirable high carbon dioxide reduction potential.

Methods

In the bioliq process, lignocellulosic biomass is first liquefied by fast pyrolysis in distributed regional plants to produce an energy-dense intermediate suitable for economic transport over long distances. Slurries of pyrolysis condensates and char, also referred to as biosyncrude, are transported to a large central gasification and synthesis plant. The bioslurry is preheated and pumped into a pressurized entrained flow gasifier, atomized with technical oxygen, and converted at > 1,200°C to an almost tar-free, low-methane syngas.

Results

Syngas - a mixture of CO and H_2 - is a well-known versatile intermediate for the selectively catalyzed production of various base chemicals or synthetic fuels. At KIT, a pilot plant has been constructed together with industrial partners to demonstrate the process chain in representative scale. The process data obtained will allow for process scale-up and reliable cost estimates. In addition, practical experience is gained.

Conclusions

The paper describes the background, principal technical concepts, and actual development status of the bioliq process. It is considered to have the potential for worldwide application in large scale since any kind of dry biomass can be used as feedstock. Thus, a significant contribution to a sustainable future energy supply could be achieved.

BACKGROUND

Only 200 years ago, the energy supply of a one billion world population depended entirely on renewables. The main energy source was firewood for residential heating, cooking, and lighting, as well as serving for high-temperature processes like iron ore reduction, burning bricks and tiles, or glass melting, etc. A complementary energy contribution was mechanical energy from hydropower for hammer mills or wind energy for windmills and sailing ships. Not to forget that the main power source for human activities carried out by working animals and human workers has been fuelled by biomass. Large energy plantations in the form of grassland and arable land (e.g., for grass, hay, oat, etc.) were devoted to 'transportation fuel' production for horses, donkeys, camels, etc.

A well-established organic chemical industry based on various biomasses also existed until about a century ago. Examples are the coproducts from thermochemical charcoal production like tar and pitch, e.g., as a glue for ship construction, wood preservatives, turpentine, 'wood spirit' (methanol), or 'wood vinegar' (acetic acid), etc. or biochemical wine and beer production by sugar and starch fermentation. It took many decades of development efforts until the major organic chemicals could be manufactured by cheaper synthetic processes from coal, crude oil, or natural gas.

Mid-2011, a world population of 7 billion people consumes around 13 Gtoe/a of primary energy [1]. The world primary energy mix consists of ca. 80% fossil fuels and ca. 10% bioenergy as shown in Figure 1. Towards the end of the century, an increase of the world population to a maximum of almost 10 billion is expected in combination with a doubling of the energy consumption to about 25 Gtoe/a. This corresponds to an average energy consumption of 3.4 kW(th)/capita or about two-thirds of the present per capita consumption in the European Union (EU 27). The economic growth takes place in the highly populated and rapidly growing and developing nations mainly in China, India, Indonesia, the neighboring South East Asia region, and in South America, e.g., Brazil, and comprises more than half of the future world population.

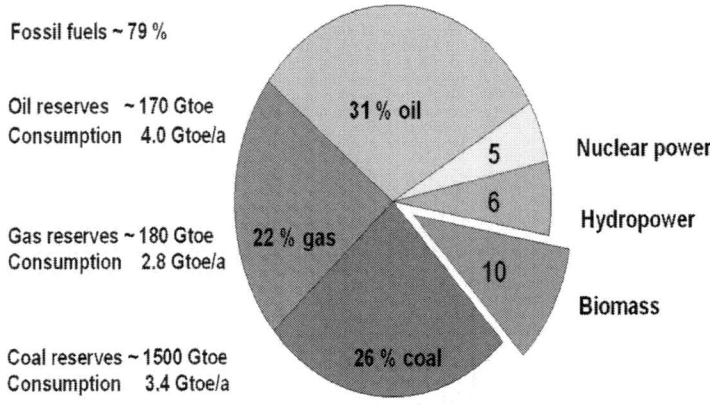

Fossil fuels ~ 79 %

Oil reserves ~ 170 Gtoe
Consumption 4.0 Gtoe/a

Gas reserves ~ 180 Gtoe
Consumption 2.8 Gtoe/a

Coal reserves ~ 1500 Gtoe
Consumption 3.4 Gtoe/a

31 % oil

22 % gas

26 % coal

5 Nuclear power

6 Hydropower

10 Biomass

Doubling from ca. 13 today to ca. 25 Gtoe/a in ca. 50-100 a
1 toe (tonne oil equivalent) = 44 GJ; 1 kW(th) = 0.375 kW(el) hydropower conversion

Figure 1: World primary energy mix 2010.

If the high fossil fuel share of *ca.* 80% would be maintained in the future energy mix, the proven and economically recoverable overall coal, oil, and gas reserves of almost 2 Ttoe [1] known in 2010 will be depleted in about a century as a continuation of the present consumption rate: first the oil in 43 years, then the gas in 62 years, and the larger coal reserves at the end in almost 400 years. However, coal will be consumed much faster when it has to take over the large oil and gas share. Together with a doubling of the energy consumption, the realistic, dynamic lifetime shrinks to a little more than 100 years. In this scenario, the present CO_2 content of 386 *v/v* in the atmosphere will about to double and cause global warming of several kelvin with rising sea levels and more frequent weather excursions.

To gradually replace the dwindling fossil fuels in the course of this century, renewable direct (photovoltaics and solar thermal) and indirect (hydropower, wind energy, and bioenergy) solar energies and quasi-inexhaustible energy sources like nuclear breeder and fusion reactors as well as some smaller contributions from geothermal and tidal energies must therefore urgently be developed to commercial maturity. The inevitable switchover of our energy supply from the finite fossil to renewable and - from a human point of view - quasi-inexhaustible energy sources requires much financial effort, time, and innovative ideas and will heavily strain human and material resources. Development and market introduction must be achieved in due time to avoid armed conflicts in case of a shortening or breakdown of energy supply. This task belongs to the major challenges of our century. Biomass must and can contribute an indispensible and significant part to a sustainable future energy supply, but with present-day technologies, it can by no means serve all energy needs of mankind. High priority has to be given to technology research and development for the inevitable exploitation of biomass as the only renewable carbon source for organic chemicals and fuels. Bioenergy is an inevitable by-product of the increasingly important biocarbon utilization.

Biomass Potential

Biomass Growth

Only about half of the 175 trillion kW(th) of solar radiation incident on the outer atmosphere of the earth arrives directly at the earth's surface, and only 0.11% of this surface energy is converted by photosynthesis to about 170 Gt/a of dry biomass (higher heating value (HHV), 5 kWh/kg), equivalent to 70 Gtoe/a of bioenergy (HHV oil, 12 kWh/kg). About 65% or 45 Gtoe are generated on land, and 35% or 25 Gtoe, in the oceans. At present, there are only speculations on how a significant fraction of the ocean biomass can be exploited, e.g., by biochemical processes in salty seawater.

About 29% of the 510-million-km^2 earth surface is land. Of the 148-million-km^2 land surface area, almost 40% is unfertile desert (too dry), tundra (too cold), or covered with ice. The large deserts of the earth extend around the tropic at latitudes of 23° north and south and separate the fertile tropical zone from the subtropical and temperate zones. About half of the about 90-million-km^2 fertile global land areas are forests; the rest of *ca.* 45 million km^2 are farmland (*ca.* 15-million-km^2 arable land plus grassland), savanna, and settlement area [2,3].

The average global upgrowth on fertile land is *ca.* 1.2 kg of dry biomass or 6 kWh(th)/m^2/year, with a large regional scatter of at least half an order of magnitude. Harvest expectations for plantations are 2 kg of dry biomass (containing *ca.* 1 kg of carbon) per m^2 and year.

Biomass combustion for electricity generation with an optimistic 45% efficiency would yield about 0.3 to 0.5 Watt(el)/m^2. Commercial photovoltaic cells are almost two orders of magnitude more efficient. Yet today, photovoltaics are still more expensive than biomass cultivation and harvest plus final combustion in conventional biomass-fired power stations.

Essential for optimal plant growth are suitable soils, temperatures, sufficient water, and fertilizer supply during the right time. C3-plants are typical for temperate climates and need about 400 kg of water transpiration via their leaves to generate 1 kg of dry biomass. C4-plants, typical for tropical and subtropical climates, need only about half. With the average rainfall on earth of roughly 700 mm/a and suitable temperatures and soil fertility, a maximum biomass harvest of about 2.5 kg/m^2(25 t/ha) can be expected for C3-plants in temperate climates; with C4-plants in tropical regions without winter season, up to 50 t/ha may be possible. Such optimum harvests may be obtained in energy plantations with irrigation and two harvests or more per year. The present world average harvests are only about half of the possible maximum. There is doubt if an optimum P-fertilization can still be provided in the future without ash recycle. In particular for large-scale biomass conversion plants, recovery of phosphorous and other minerals is a must.

In the EU 27 with 1,160,000-km^2 arable land, a part of 6.7% is already set aside [4] to avoid an expensive overproduction of food. If optimum agricultural technologies are applied in all EU countries, up to 20% of the arable land or even more can be set aside or used for biomass plantations. Assuming an average harvest of 20 t/a of dry biomass/ha, a total harvest of almost 0.2 Gtoe/a (containing 0.25 Gt of biocarbon) might be realized in few decades. Even without the residues from agriculture and forestry in comparable amounts, this is sufficient for a sustainable supply of both organic chemistry and aviation fuel production. Most studies estimate that the bioenergy contribution in the EU will increase to more than 10% after 2020 and to more than 20% on the longer term [5]. In the latter case, the major part must then be supplied from energy plantations. Different from agricultural or forest residues, all direct and indirect costs of plant cultivation must then be charged to the bioenergy. The advantages of energy plantations in tropical regions are clearly visible in Table 1 from the two to three times higher hectare yields for liquid biofuels.

Table 1: Potential biofuel yields per hectare in temperate and tropical climates

Climate	Crop/country	Crop residue	Biofuel type	Yield (t/ha)	Diesel equivalent (sum; t/ha)
Temperate climate	Sugar beet; Germany	Sugar	Ethanol	4	3
	Rape seed; Germany, USA	Oilseed; straw	FAME; FT diesel	1.2; 0.5	1.7
Tropical climate	Palm oil; Malaysia	Oilfruits; palm waste	FAME; FT diesel	6; 2	8
	Sugar cane; Brazil	Sugar; bagasse	Ethanol; FT diesel	6; 2	6.5

FT, Fischer-Tropsch.

Dahmen et al. Energy, Sustainability and Society 2012 2:3   doi:10.1186/2192-0567-2-3

Competitive Biomass Use and Harvest Limits

The most abundant constituent of terrestrial plants is lignocellulose with more than 90 wt.%, the water-insoluble polymeric construction material of the cell walls. Dry lignocellulose is composed of about 50 wt.% cellulose fibers, wrapped up and protected in sheets of ca. 25 wt.% hemicellulose andca. 25 wt.% lignin. Any large-scale biomass use must rely on this most abundant biocarbon material. Starch, sugar, oil, or protein in food crops are far less abundant, and their use as human or animal food or feed has the highest priority.

It is an important issue how much of the terrestrial biomass upgrowth of ca. 45 Gtoe/a (ca. 110 Gt/a of dry biomass) is possible and desirable to harvest. Almost half of the global land biomass upgrowth consists of the annually falling leaves and needles in the forests [2], above all in the tropical rain forests. They can neither be collected with reasonable effort nor used since their high mineral content makes them indispensible as an on-site fertilizer. The biomass harvest is further diminished by harvest losses and residues like tree stocks, roots, plus stubble of cereals, etc. left on-site, as well as by storage losses of wet biomass via biological degradation at more than ca. 15 wt.% water content.

Limits for a secure prevention of overexploitation are not reliably known. For the EU 27 with an actual gross inland energy consumption of 1.9 Gtoe/a, the bioenergy contribution of 4% is estimated to increase sustainably to almost 15% or 300 Mtoe/a of the energy consumption expected for 2030[4,5]. A rather optimistic potential future scenario is presented in Table 2: about a quarter of all terrestrial biomass upgrowth or 11 Mtoe/a can be harvested and used sustainably for all biocarbon and bioenergy applications. This is almost three times the present use and probably not far from a sustainable upper limit.

Table 2: Biomass utilization scenario compared to the present use

Biocarbon/bioenergy use	Year/population	
	2011/7 billion (Gtoe/a[a])	**2050+/10 billion (Gtoe/a[a])**
Biocarbon use for		
1. Human plus domestic animal food and feed; food harvest residues (e.g., straw)	*ca.* 2 < 0.2	2.5 0.5
2. Construction wood (timber)	0.5	> 1
3. Plantations for special organic raw materials (cellulose fiber, cotton, pulp and paper, caoutchouk, oilseed for detergents, etc.)	*ca.* 0.2	1
4. Synthetic organic chemistry by bio- and thermochemical routes with cogeneration of energy	< 0.1	1
Bioenergy use for		
5. Traditional firewood combustion, etc.	1	1
6. Energy for high-temperature processes (cement, lime, bricks, ceramic production, etc.)	< 0.1	0.5
7. Ore reductant (mainly iron ore)	< 0.1	0.5

8. Aviation, ship, and special car fuels (assuming 50% BTL energy conversion efficiency)	< 0.1	2
9. CHP in remote areas	< 0.1	1
Total biomass consumption (1 to 9)	ca. 4	ca. 11

[a]1 Gtoe is ca. 2.4 t of lignocellulose free of water and ash. BTL, biomass to liquid; CHP, combined heat and power.

Dahmen et al. Energy, Sustainability and Society 2012 2:3 doi:10.1186/2192-0567-2-3

Human and animal food production is indispensible and is the first priority. The second priority is stem wood utilization as the still dominant organic construction material (timber) as well as the production of organic raw materials like cellulose fibers from wood or cotton, caoutchouc, or extracts like flavors, drugs, dyes, etc. In the future, when the fossil hydrocarbon reserves become too expensive or exhausted, all applications utilizing biofeedstock as the only renewable carbon resource will gradually gain higher priorities. Direct biomass combustion for heat, power, and electricity generation today still enjoys high priority to fight global warming because combustion is in most cases economically more favorable than using lignocellulosic biocarbon via gasification or fermentation as the only renewable carbon raw material for organic chemicals and fuels [6], yet this is only an intermediate situation as long as fossil fuels are still available. All other renewable energy sources produce heat or electricity directly but no carbon. Moreover, thermochemical biomass conversions also generate energy as an inevitable couple-product in the form of reaction heat and sensible heat of the reaction products. In future biorefineries, the cogeneration of energy will be normal and used to rise high-pressure steam, power, or electricity, mainly to supply the own self-sustained process and to export any potential surplus.

The amount of carbon needed for organic chemistry is only about 4% compared to the amount which would be required for global energy supply via combustion. The 2050+ scenario in Table 2 shows that even with a massive increase of biomass use, only ca. 6 Gtoe/a or about a quarter of the future global primary energy demand can be covered by biomass. Supply of the much smaller carbon fraction for organic chemistry does not cause much problem.

In some cases, carbon-based energy production is difficult to replace, in particular in the transportation sector. Even if all road transport can be electrified, a significant amount of liquid hydrocarbon transportation fuel will be needed at least for aviation, probably also for ship transport and for car, bus, and truck transports in remote areas. Producing 1 Gtoe/a of biosynfuel for these special applications requires *ca.* 2 Gtoe/a of lignocellulose as a raw material, a significant share of the total bioenergy harvest. Carbon materials are also needed for iron ore reduction, *ca.* 0.5 Gtoe/a of charcoal might be a reasonable estimate toward the end of the century. In steel and glass production, as a part of the high-temperature process, heat can be supplied in the form of electricity. Corresponding electro-technologies do not exist for the present global cement production of 2.2 Gt/a or for bricks, lime, ceramics, tiles, etc. production. The traditional direct biomass combustion for home heating and cooking is assumed to continue at the present level together with some additional CHP applications.

Wood and Straw

The terms wood and straw are used here only as synonyms for slow- and fast-growing lignocellulosic biomass with low (< 3 wt.%) or higher ash content, respectively. Wood without bark is a relatively clean biofuel with a typical ash content of 1 wt.% or below. Fast-growing biomass from agriculture like cereal straw, grass, hay, etc. has an ash content between 5 and 10 wt.%, rice straw even 15 to 20 wt.%. Wood ash contains much CaO, straw ash about half SiO_2 with much K and Cl. These and other inorganic constituents are needed as part of the biocatalyst systems, which are responsible for a faster metabolism. Higher ash and heteroatom (e.g., N, S) contents are therefore also typical for the faster growing aquatic plants and for active animals. This is simultaneously a hint to higher fertilizer costs for plant cultivation.

Combustion and gasification technologies for low-quality biofuels with much ash are not well developed. Special technical problems with straw and straw-like materials in thermochemical processes are:

- Potassium can reduce the ash melting point down to less than 700°C (eutectics!). Sticky ash during either combustion or gasification increases the risk of reactor slagging.
- Chlorine is released mainly as HCl, causing corrosion in gas cleaning facilities, poisoning catalysts, and potentially inducing the formation of toxic polychlorinated dibenzodioxins or furans due to unsuitable combustion conditions.
- Volatility of alkali chlorides (in particular of KCl) at temperatures above 600°C can cause deposits, plugging, and corrosion in gas cleaning systems.
- Ash and volatile organic carbon impurities can create problems during co-combustion or co-gasification. Fuel nitrogen in the form of proteins is partly converted to NO.

High nitrogen contents are mainly converted to N_2 and must be compensated by expensive N-fertilizers. Thermochemical processing is therefore not suited for protein-rich biomass ($N = 16\%$ of the protein weight) with a N content above about 3 wt.%.

The elementary CHO composition of dry, ash-, and heteroatom-free lignocellulose in different biofeedstock is almost the same and well represented by $C_1 H_{1.45} O_{0.66}$. A reasonable sum formula with integer atom numbers is $C_6 H(H_2O)_4 \triangleq C_1 H_{1.5} O_{0.67}$ or $C_9 H(H_2O)_6 \triangleq C_1 H_{1.44} O_{0.67}$. An even simpler and still reasonable sum formula is $C_3 (H_2O)_2 \triangleq C_1 H_{1.33} O_{0.67}$, a 1:1 formal mix of carbon and water in weight. The HHV of dry, ash-free lignocellulose is *ca.* 20% higher than a simple 1:1 wt.% carbon/water mix. However, this simple picture is useful for quick stoichiometric estimates. In comparison to glucose, as the primary organic product of photosynthesis, the sum formula $C_6 H_8 O_4$ is also used. To represent real biomass, some ash and moisture must be added to the lignocellulose. Heteroatoms like N or S can, in most cases, be neglected to a first approximation, except in protein-rich biomass (nitrogen in protein, *ca.* 16 wt.%). The sulfur content usually is rather low, about an order of magnitude compared to coal.

Basic Concept Considerations

Biomass utilization will increase in the future not only due to the growing food consumption for a larger population, but also due to the extension of old and new bioenergies and especially biocarbon applications, required to gradually substitute fossil carbon and hydrogen. Our technology selection criteria for biomass refining processes have been based on general and global considerations [7], not on regional particularities.

Conclusions from the above-mentioned aspects

- *Bioenergy generation at the expense of poor food supply must be strictly prevented.* Direct use of biomaterials with complex chemical and physical structures like wood as construction material, cotton, caoutchouc, etc. has also a higher priority than combustion.

- Use of *biomass as the only renewable carbon resource* for valuable organic materials, chemicals, and fuels has a higher priority than the generation of bioenergy via combustion.

- At present, the most urgent task is the development of biomass conversion technologies for *liquid transportation fuels* [8] to decrease our oil dependency. Supply security is the most important aspect on the short term. Politically motivated brief shortages of oil supply or extremely high prices of crude oil can cause a serious breakdown of the world economy with a risk of armed conflicts.

- *Biorefineries* are an inevitable long-term development task for the production of all types of carbon materials from biomass. Biomass conversion to organic chemicals or to liquid transportation fuels requires several chemical reactions in succession. Energy is an inevitable couple and side product. In comparison to zero feed cost, biomass-to-liquid (BTL) processes require more technical effort than in an oil refinery. This results in a *lower overall energy recovery* in the final product and *higher manufacturing costs*.

- *Biocarbon supply is limited.* A secure and sustainable upper supply limit for biomass is not reliably known. An optimistic upper limit estimate after 2050 assumes that about a quarter of all land biomass can be exploited for everything from food to combustion (see Table 2). The present global bioenergy contribution of > 1 Gtoe/a can probably be increased sustainably to *ca.* 5 to 6 Gtoe/a, a factor of *ca.* 5. When bioenergy consumption approaches this upper limit, not only the biomass prices will increase, but also the food prices due to the mutually competitive land use. Because of the unknown bio-production limits, there is a high risk of overexploitation with a potential breakdown of bio-production for decades or centuries, as already experienced with deforestation in some Mediterranean regions.

- Without fossil carbon, some *new or renewed bioenergy applications* will emerge, in cases where carbon is needed and a direct use of renewable electrical or mechanical power is unsuited or too expensive. Examples are:

- For iron ore reduction, generation of either charcoal or CO or ($CO + H_2$) mixtures via pyrolysis is a renewed old technology.

- Heat generation for high-temperature processes for cement, bricks, lime, etc. production.

- *Conventional biomass combustion* for residential heating and cooking is assumed to continue at about the present level and is probably complemented by additional CHP-plants for simultaneous heat and electricity generation in remote areas.

- In a few decades, road or car electrification will probably complement the electrified rail.

However, the convenient liquid hydrocarbons are hard to replace as *aviation fuels* - eventually also as *ship fuels* and for the still remaining fraction of car, truck, and bus fuels. In the course of the century, the biomass demand for these conventional and new synthetic transportation fuels, tailored for new or optimized engine types, might probably become higher than that for organic chemicals. The production technology for biosynfuels and organic chemicals do not differ principally. However, liquid organic fuels belong to the cheapest organic chemicals.

- *Bioenergy* can sustainably cover probably up to a quarter of the future global primary energy demand. The crude estimate in Table 2 indicates a maximum bioenergy contribution of *ca.* 6 Gtoe/a including the couple-product energy from chemical conversions. During thermochemical biocarbon conversion, about half of the initial bioenergy on the average is typically liberated in exothermal reactions in the forms of reaction energy and sensible heat. Recovery and conversion of half of this energy, e.g., in high-pressure steam or electricity, make use of about a quarter of the initial bioenergy as a couple-product.

Biorefineries

A biorefinery [9] is a flexible coherent system of physical and chemical facilities for the conversion of all types of biomass into more valuable organic materials, chemicals, and fuels; heat, power, and electricity are inevitable couple and side products from exothermal chemical reactions. This network for the simultaneous cogeneration of carbon materials and energy is nothing new, but the normal situation in any integrated multistep organic chemistry is complex. Biorefineries are the organic chemical industry of the future and use biomass as a carbon raw material. Energy, especially in the form of heat or high-pressure steam, can be consumed on-site to generate a self-sustained process; an energy surplus is usually exported as electricity and credited to the main products. Biorefineries can be classified according to the main conversion process into:

- Physicochemical - e.g., pulp and paper mills, sugar mills, corn mills, fatty acid methyl ester plants, etc.
- Biochemical - low-temperature wet processes with high selectivity (ethanol, butanol, biogas, etc.)
- Thermochemical - high-temperature dry processes proceed usually via syngas, e.g., BTL technology.

Additional classification aspects - without considering educts and products - are the main intermediate(s) (platform chemicals), which are suited for mutual exchange between plants. This script reports about a development work for the 'backbone' conversion steps of a thermochemical biorefinery: conversion of the abundant lignocellulose via biosyngas - a mix of CO and H_2 - as a versatile intermediate to H_2, CH_4, CH_3OH [10,11], dimethyl ether (DME), Fischer-Tropsch (FT) hydrocarbons, [12] or other carbon products, using highly selective catalysts at specified temperatures and higher pressures. Most synthesis steps are known since almost a century and are practiced already on the technical scale [13,14] based on coal and natural gas as feedstock known as coal-to-liquid (CTL) and gas-to-liquid (GTL) processes. Examples are the CTL plants operated by Sasol in South Africa or the Shell GTL plants in Malaysia or Qatar. The development of BTL is not completed but, to a large extent, can rely on the old coal conversion technologies in an improved or modified form. Major development work is needed especially for the front-end steps to prepare a clean syngas from various biofeedstock types. After generation of a clean syngas with the desired H_2/CO ratio, the BTL technology is comparable with the practiced CTL and GTL technologies since it does not make a difference if the syngas has been produced from coal, oil, natural gas, biomass, or organic waste. Syngas or C_1 chemistry in general is based on a well-known technology [13,15]. This is why the actual work at the Karlsruhe Institute of Technology (KIT) has been focused mainly on the front-end BTL steps.

Selection of Gasifiers for Biomass

Gasifier Types

The typical gasifier types [16] for coal shown in Figure 2 can also be used for lignocellulosic biomass after special preparation [17]. Suitable feed particle size and gasification reaction times decrease from about 0.1 m and more than 10^3 s for fixed bed gasifiers, via *ca.* 1 cm and 10^2 to 10^3 s for fluidized bed gasifiers, down to $\leq$ 0.1-mm fuel powders, which react in one or few seconds in an entrained flow (EF) gasifier flame. Short reactor residence times and higher pressures result in smaller and more economic reactors with a higher throughput.

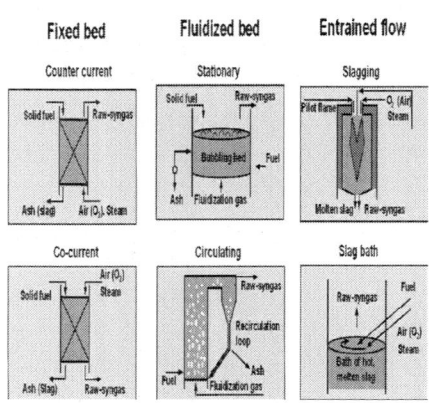

Figure 2: Gasifier types suited for coal and biomass.

Fixed and fluidized bed gasifiers operate with solid ash at temperatures below 1,000°C. Low-melting straw ash can become sticky already at 700°C and can create problems by bed agglomeration. Raw syngas from fixed and fluidized beds contains tar and methane because of the low gasification temperatures; especially, the syngas from updraft gasifiers is contaminated with much dirty pyrolysis gas. Syngas applications for combustion can tolerate high methane contents and require less gas cleaning efforts. EF gasifiers operate above the ash melting point at > 1,000°C and generate a practically tar-free, low-methane raw syngas.

Because of the higher temperatures in an EF gasifier, a cleaner syngas is obtained at the expense of more oxygen or air consumption and correspondingly lower cold gas efficiency. However, this is at least partly compensated for by the low methane content, which would otherwise reduce the $CO + H_2$syngas yield by 4% for every percent of CH_4: $CO + 3H_2 \rightleftarrows CH_4 + H_2O$.

Synthesis reactions with syngas are exothermal and generate larger molecules, except the CO-shift reaction to H_2. Equilibrium yields and kinetics are therefore improved by higher pressures, usually in the range of 10 to 100 bar. Slagging EF gasifiers can be designed for higher pressures up to 100 bar and allow for higher and more economic capacities up to 1 GW(th) or more. Another contribution to synthesis economy is the use of pure oxygen as a gasification agent to avoid syngas dilution to about half with N_2 from air.

Selection of the GSP-Type Gasifier

Key step of the KIT bioliq process [18-28] is an oxygen-blown, slagging EF gasifier operated at high pressure above the downstream synthesis pressure up to *ca.* 80 bar and at gasification temperatures ≥ 1,200°C above the ash melting point to generate a tar-free, low-methane syngas from liquefied biomass. The general advantages of slagging highly pressurized EF gasifiers (PEF) [16] can be briefly summarized as follows:

- Tar-free syngas with low CH_4 contents
- High reaction pressures and temperatures possible
- High (> 99%) carbon conversion
- High capacities (≥ 1 GW(th)) possible

- High feed flexibility; according to the high conversion temperatures, the gasifier is a 'guzzler.' With a modified burner head biooils, bioslurries and biochar powder can be gasified.

Precondition for EF gasification is the conversion of a solid feedstock to a gas, liquid, slurry, or paste, which can easily be transferred by a compressor or pump into the pressurized gasifier chamber. Any organic feed stream with a HHV > 10 MJ/kg, which can be pumped and atomized in a special nozzle with pressurized oxygen as gasification and atomization agent, is suitable. At moderate pressures, a dense stream of fine char or coal powders can also be fed pneumatically from a pressurized fluid bed with an inert gas stream [29], similar to pulverized, coal-fired burners in power stations. At increased pressures, the powder transport density remains nearly the same, and more transport gas is required.

At a sufficiently high gasification temperature, slag with oil- or honey-like viscosity drains down at the inner wall, drops into a water bath below the gasification chamber for cooling, and is removed periodically via a lock. The large volume flow of hot syngas through the lower central opening of the membrane screen vessel causes a certain pressure drop, which is measured. A higher pressure drop indicates a narrowing of the exit hole by highly viscous slag. This automatically increases the oxygen flow and thus the gasifier temperature until the slag is molten and drained. Additives or slag recycle can be helpful to maintain a sufficiently low slag melting temperature and thus to limit oxygen consumption at a still sufficiently high gasification rate.

The outer, pressure-resistant, mild steel shell behind the membrane wall attains only about 250°C cooling water temperature, which does not affect the mechanical stability. The special advantages of a Gaskombinat Schwarze Pumpe (GSP)-type PEF gasifier are briefly summarized as follows:

- The membrane wall with SiC refractory permits the gasification of fuels with much ash and corrosive salts, as is typical for straw and straw-like, fast-growing biomass.
- The relatively thin membrane wall plus slag layer has a low heat capacity and allows frequent and fast start-up and sudden shutdown procedures without damaging the gasifier, e.g., in case of an accidental feed interruption.

- The membrane wall design with protecting slag layer guarantees long service life for many years, as has been shown in more than 20 years of operation with various feeds in the 130-MW(th) GSP gasifier at 'Schwarze Pumpe', East Germany [29,30].

A disadvantage is the high heat loss of 100 to 200 kW/m^2 through the thin slag and SiC layer at the membrane wall, depending on the thickness and composition of the slag layer. In small pilot gasifiers with only few megawatt power, the large surface-to-volume ratio causes a considerable heat loss of several 10% and requires careful data correction for scale-up considerations. In large commercial gasifiers with a capacity of several 100 MW(th), the losses via the membrane screen drop to below 1% and become negligible. This shows that the GSP gasifier is not recommendable for small-scale plants.

The GSP-type (gasification complex 'black pump') has been developed in the 1970s in the Deutsches Brennstoff Institut (DBI), Freiberg, East Germany, for the salt (NaCl)-containing lignite from Central Germany, which poses corrosion problems with alkali chlorides similar to KCl-containing slag from fast-growing biomass [29,31-33]. Figure 3 shows the simplified GSP gasifier design. The internal cooling screen is a gastight, welded membrane wall of cooling pipes with a thin inner SiC liner, particularly suited for low-quality biomass with much low melting slag from KCl-containing ash. The pipes are cooled with pressurized water at 200°C to 300°C. A thin, ca. centimeter-thick, viscous slag layer covers and protects the inner surface of the membrane wall from corrosion and erosion. Only a small slag fraction of a few percent escapes in the form of tiny, sticky droplets with the raw syngas. In 1996, an experienced development personnel designed and built an improved 3- to 5-MW(th) GSP pilot gasifier in Freiberg to test the hazardous waste conversion process of Noell Company [34]. Experience with the GSP gasifier is the sound basis of the KIT concept. The KIT bioslurry gasification concept has been verified and investigated in this pilot gasifier in four gasification campaigns in year 2002, 2003, 2004, and 2005 in cooperation with Future Energy, today Siemens Fuel Gasification Technologies.

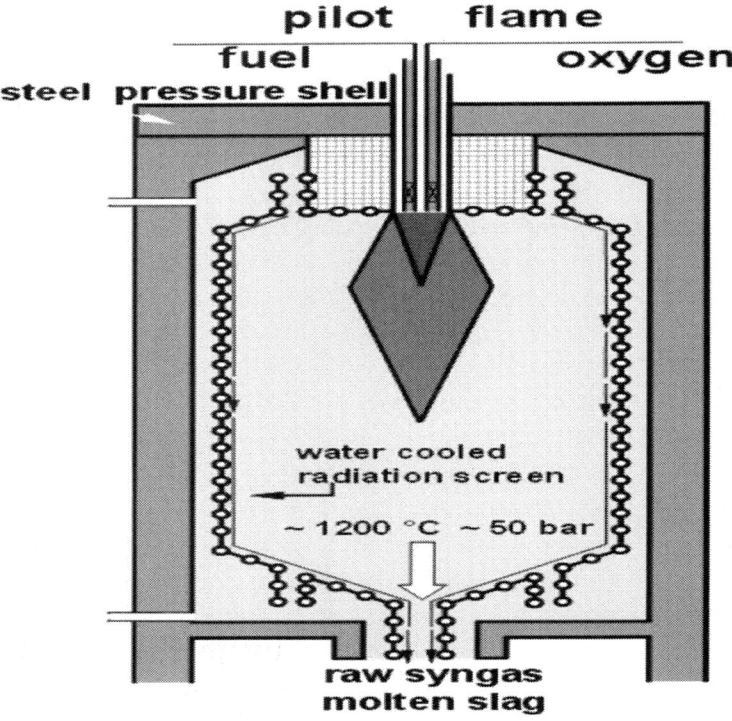

Figure 3: Scheme of a PEF gasifier with cooling screen.

At KIT, a 5-MW(th) pilot gasifier with a cooled membrane wall for a maximum of 80-bar pressure is presently being constructed as a part of the bioliq pilot facility for the production of synthetic biofuels from biomass. Substantial financial support has been granted by FNR (German Ministry of Agriculture). Responsible for the design, erection, and commissioning of the PEF pilot gasifier with a membrane wall is Lurgi AG Company, Frankfurt; start-up is expected in 2012.

Several companies have recognized the advantages of slagging PEF gasifiers for biomass conversion to syngas; Table 3 gives a brief overview. The main difference between these process variants are the biomass pretreatment steps. Pretreatment for PEF gasifiers requires more technical effort than that for fixed or fluidized bed gasifiers.

Table 3: BTL developments using PEF gasifiers

Company/country	Gasifier feed	Gasification conditions	Biomass pretreatment
Schwarze Pumpe/ Germany [29-33]	Diverse liquids, slurries from waste and lignite	26 bar, 1,200°C to 1,600°C, GSP-type, 130 MW(th)	Diverse lignite, organic waste
Choren/ Germany[5,91,92]	Hot pyrolysis vapors, char powder for chemical quench	4 to 5 bar, > 1,400°C, char quench to 900°C	Auto-thermal pyrolysis on-site at gasifier pressure
Chemrec/Sweden [93,94]	Concentrated black liquor	ca. 40 bar, ca. 950°C	Integrated into the on-site pulp mill
KIT, bioliq/ Germany [18-27]	Any bioslurry or paste from biooil plus char	up to 80 bar, ca. 1,200°C	FP at 500°C on- or off-site; any type of biomass liquefaction
ECN/The Netherlands [95,96]	Pulverized char from torrefaction	ca. 40 bar, ca. 1,200°C	Torrefaction (≤ 300°C pyrolysis on- or off-site)
BioTFueL/France	Pulverized char from torrefaction	Uhde Prenflow™ gasifier, 15 MW(th)	Torrefaction

Outline of the Bioliq® Process

The bioslurry-based BTL process of KIT called bioliq is described in more detail in the works of Henrich and colleagues [18-27]. The simplified process scheme in Figure 4 gives an overview.

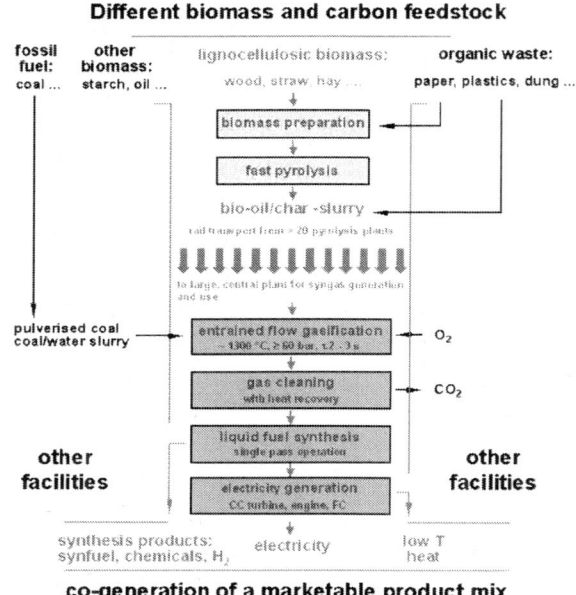

Different biomass and carbon feedstock

co-generation of a marketable product mix

Figure 4: Block flow diagram of the bioliq® process.

Biomass Preparation and Fast Pyrolysis

Sufficiently dry lignocellulosic biomass like wood or straw below *ca.* 15 wt.% moisture can be stored without biological degradation. The dry biomaterials are diminuted in two steps into small particles of < 3 mm in size. The energy required for diminution is reduced at lower moisture.

Biomass particles with a characteristic length of < 0.5 mm (sphere diameter, < 3 mm; cylinders, < 2 mm; plates, < 1 mm) which are equivalent to a specific surface of > 2,000-m^2/m^3 biomass volume are mixed at atmospheric pressure and at temperatures of *ca.* 500°C under exclusion of air with an excess of a hot, grainy heat carrier like sand or stainless steel (SS) balls [27,35]. In principle, any fast pyrolysis (FP) reactor type [36] can be applied. At KIT, an FP system with a twin-screw mixer reactor is being developed, based on the Lurgi-Ruhrgas system. The thermal decomposition of biomass and the condensation of organic tar vapors and reaction water vapors take place in the course of one or few seconds. High condensate yields of 45 to 75 wt.% are coupled with low char and gas yields; this is typical for FP. The char contains all ash; the solids' yield depends on feedstock and operating conditions and is in the range between *ca.* 10 and 35 wt.%. The pyrolysis gases contain CO and CO_2 as main components in amounts between 30 and 55 vol.%; methane, hydrogen, and hydrocarbons up to C_5 amount to *ca.* 10 vol.%. The heating value of the pyrolysis gas is about 9 MJ/kg. The total energy content of the FP gas corresponds to about 10% of the initial biomass HHV and is sufficient to supply the thermal energy for a well-designed FP reactor.

Production of Bioslurries

FP char contains about 20% to 40% of the initial bioenergy; the condensate (biooil), 70% to 50%, and together, about 90%. If only the biooil is used for gasification without the char, about one-third of the bioenergy would not be accessible for syngas generation. Therefore, the pyrolysis char powder is mixed into the biooil to generate a dense slurry or paste with a density of about 1,200 kg/m³ and a HHV from 18 to 25 GJ/m³ which corresponds to one-half up to two-thirds of the volumetric energy density of heating oil (HHV 36 GJ/m³) [37-39].

There are many good reasons for bioslurry production: A single pyrolysis product with high energy density eases handling, storage, and transport; a free-flowing bioslurry can be conveniently pumped with little effort into highly pressurized gasifiers. Even low-quality biooils which are prone to phase separation and are contaminated with char and ash are still suited for bioslurry or paste preparation. The fine, porous pyrolysis char powders from FP are very sensitive to self-ignition (self-ignition temperature is typically > 115°C), and fine, airborne, char dust particles can penetrate breathing masks. Pulverized biochar usually is pelletized for safety and handling reasons; slurries provide a much safer way of char handling.

Pef Gasification of Bioslurries

Not only bioslurries and pastes, but also other dense forms like char crumbs soaked with tar or pelletized biochar can be transported in silo wagons with the electrified rail from several dozens of regional pyrolysis plants into a large, central biosynfuel plant for syngas generation and use. PEF gasification is a complex technology, and a large scale is required due to economy-of-scale reasons. A suitable menu of bioslurries is preheated with waste heat from the process to reduce the viscosity and mixed in large vessels to obtain the desired composition and is then further homogenized in robust colloid mixers [40] during feeding. The preheated slurry is transferred with screw or plunger pumps into a highly pressurized PEF gasifier and pneumatically atomized in a special nozzle system with pure oxygen. Gasification to a tar-free, low-methane syngas proceeds in 3 to 4 s in a downward flame [16] at ≥ 1,200°C above the ash melting point and at pressures up to 100 bar. In a GSP-type gasifier[29], a viscous, honey-like, *ca.* 1-cm-thick slag layer drains down at the inner surface of a cooled membrane wall and protects the gasifier from erosion and corrosion. Gasifier compatibility with the corrosive biomass ashes is an essential characteristic. The high pressure slightly above the downstream syntheses pressure eliminates the high investment and operating costs for an intermediate syngas compressor station and reduces the PEF reactor size. The pilot gasifier currently erected at KIT can be operated at pressures up to 80 bar.

Cleaning and Conditioning of the Raw Syngas

Syngas is a 'platform chemical' which can be used for many different purposes: (1) combustion for a high-temperature process of heat generation, (2) as fuel gas in IGCC power stations, or (3) in small CHP plants with stationary gas motors or turbines. Moderate gas cleaning is required for these applications. Practically, no syngas cleaning is needed for iron ore reduction. A very efficient raw syngas cleaning and conditioning section is needed prior to a catalyzed chemical synthesis [41]. Slag and soot particles, tars, alkali salts, and gaseous S-, N-, and Cl-containing impurities like H_2S, COS, CS_2, NH_3, HCN, HCl, etc. have to be removed down to below the part-per-million level to prevent poisoning of the highly selective but sensitive catalysts. The lower the catalyst temperature, the higher the selectivity, but the sensitivity to impurities is also a rule of thumb. Conventional technologies for gas cleaning are available, e.g., the well-established Rectisol process with methanol.

Most syngas reactions require an optimum H_2/CO ratio, which is usually obtained via CO conversion to H_2 with the catalyzed homogeneous shift reaction $CO + H_2O \rightleftarrows CO_2 + H_2$; a Fe/Cr catalyst is applied for the high-temperature shift at $ca.$ 400°C, and a Cu/Zn catalyst is applied for the low-temperature shift at $ca.$ 200°C; a sulfur-resistant MoS_2/Co catalyst is suited at $ca.$ 300°C. CO_2 removal downstream is possible with a number of absorbers; the conventional Rectisol process [41] removes all higher boiling impurities by absorption in cold methanol at $ca.$ -50°C; this is a well-known and very efficient but expensive technology, yet one of our objectives is to look for process variants without the necessity of an expensive CO shift. In addition, in the pilot facility of the bioliq process, a hot gas cleaning system is applied, consisting of a ceramic particle filter, a fixed bed sorption for sour gas and alkaline removal, and a catalytic reactor for the decomposition of organic (if formed) and sulfur- and nitrogen-containing compounds [42].

Syngas Use

Clean syngas with the desired H_2/CO ratio, temperature, and pressure is routed to one of the highly selective catalysts for the production of H_2, CH_4, methanol (CH_3OH), DME, CH_3OCH_3, olefins, alcohols, FT hydrocarbons, or other chemicals [43,44]. Synthesis selectivity permits a flexible switch-over into different routes of organic chemistry. Except H_2 production via the CO-shift reaction, the synthesis of larger molecules proceeds under volume reduction, and higher pressures favor product formation at equilibrium. Because of the order increase in the product molecules, the reaction entropy $\Delta_r S$ is negative, and lower temperatures shift the equilibrium to the product side. At lower temperatures, more active catalysts are needed, which are more sensitive to trace impurities and require more efficient gas cleaning; also, a conversion of the reaction heat to power and electricity becomes less efficient.

Most synthesis reactions with syngas are highly exothermal, and efficient heat removal is the main problem of the reactor design. The major reactor types are tubular, staged, or slurry reactors with efficient coolers. The immense literature on catalytic syngas conversion is summarized in reviews [13,15], monographs [44], and handbooks [14]. Reasonable pathways to biosynfuels are the FT synthesis and the methanol route [10,13, 45]. The FT product spectrum depends on the temperature (200°C to 350°C), pressure (15 to 40 bar), reactor type, and catalyst, usually Fe or Co, and extends from gaseous CH_4 and C_2-C_5 alkanes, a C_5-C_9 gasoline, and a C_{10}-C_{20} diesel fraction of n-alkanes up to linear C_{100} waxes. Fe catalysts catalyze also the CO-shift reaction and allow operation with H_2/CO ratios below 2 in the feed gas. To increase the biosynfuel yield, the C_{25+} product waxes are catalytically converted into gasoline and diesel in a hydrocracker.

Present focus of the bioliq process is the production of gasoline via DME (boiling point (b.p.) 24°C)[46-48] as a chemical intermediate to organic chemicals and biosynfuels. Neat DME is suited as a clean and environmentally compatible diesel fuel for cold climates. For the one-step synthesis of DME in the bioliq process, a mixture of a low-temperature $Cu/ZnO/Al_2O_3$ methanol catalyst and an alumina or zeolite dehydration catalyst is used. Since the methanol catalyst also catalyzes the CO-shift reaction, a lower H_2/CO ratio of 1 or even below offers the possibility of a cheaper syngas purification train without CO shift. In addition, the high, thermodynamic DME yields at higher pressure offer the possibility of a single-pass synthesis without expensive recycle of unreacted syngas.

Based on the considerations made above, a complete BTL process chain is erected at KIT. The bioliq process will be covered on the pilot plant scale in four successive process sections, with the aim to determine design data for commercial facilities, to gain practical experience, to allow for reliable cost estimates, and for further process development and optimization. The plant consists of:

- A 2-MW(th) (0.5 t/h), pilot-scale FP of lignocellulosic materials and biosyncrude preparation
- Bioslurry PEF gasification up to 80 bar in a 5-MW(th) pilot gasifier with a membrane screen
- High-temperature, high-pressure raw syngas cleaning and conditioning, H_2/CO ratio adjustment, and CO_2 separation
- Conversion of a *ca.* 700-Nm³ synthesis sidestream to gasoline via

DME with integrated CO shift

The FP plant is already in operation; the other three plants are under construction with start-up expected in 2012. In the years to come, the present focus on biosynfuel will gradually shift to chemical products. The status of the KIT bioliq pilot plant is reported elsewhere [49,50]. A photo of the construction site is shown in Figure 5.

Figure 5: The bioliq® pilot plant construction site.

The following chapters explain the conceptual design and process fundamentals in essential details and the research and development status for the KIT bioliq process in sequence of the successive process steps. Finally, a cost estimate is presented.

Biomass Preparation for FP

Any type of dry lignocellulosic biomass can be exploited with the bioliq process. The present experimental program at KIT focuses on low-quality lignocellulosic biomass, which is rarely used and still available in larger amounts in central Europe. This amounts to about half of the cereal straw harvest which is not used and not needed to maintain soil fertility. As a crude overall estimate, it can be assumed that the average grain-to-straw ratio is about 1. The world grain harvest (wheat, maize, rice, and barley together *ca.* 90%) amounts to 2.2 Gt/a, and thus, about 1.1 Gt/a of surplus straw will be available, a significant energy equivalent of 0.4 Gtoe/a. Residues from the logwood (timber) harvest like bark, twigs, and other forest residues can contribute a comparable amount. The cost for cultivation and harvest of these bio-residues is covered by the main products.

We have checked the conventional drying, diminution, and heating processes for various biomaterials. Drying to less than 15 wt.% water content is desirable to prevent biological degradation during storage. Up to now, we have focused on a two-stage diminution of air-dry straw: first in a usual chaff cutter followed by a hammer mill to smash the several-millimeter-thick stalk nodes. Nodes come to about 5% of the straw mass and increase the heat-up time and reactor size for FP with the square of the particle size. The typical wall thickness of cereal straw is about 0.3 mm and corresponds to a specific surface of almost 7,000 m²/m³. The reciprocal specific surface is the shape-independent characteristic length of 0.15 mm. Diminution to a single-walled straw material down to about 1 cm in length is sufficient; further diminution is not desirable because it does not change the characteristic lengths, and excessive diminution creates dust problems.

We also operate a shredder for the first diminution of large pieces and a cutting mill for the second stage. The latter turned out to be suited even with dump knives. A hammer mill is also suited for the final diminution of wood chips to below 3 mm. Due to the large variety of biomaterials, there is no standard solution for optimum diminution. Drying increases the brittleness and reduces the energy consumption for diminution.

FP of Lignocellulosic Biomass

Previous Work and Conclusions

After the first oil price crisis in 1973, the development of FP of lignocellulosic biomass was pushed mainly in Canada, where huge forest resources and a low population density create a high mass potential. Conversion of wood in a simple, single step at a moderate temperature of about 500°C into a stable and clean liquid fuel called biooil was a charming idea [51,52]. The vision was to replace part of the crude oil-derived heating oil and to substitute a substantial part of the oil-derived motor fuels not only for stationary applications, but hopefully also for mobile internal combustion engines in passenger cars, busses, and trucks. Today, three decades later, no commercial biomass FP plant is in operation for 'biooil' motor fuel production. On the contrary, most of the FP pilot plants which have been designed, built, and operated for some time have been decommissioned or mothballed. Reported reasons are low oil prices, high biomass prices, poor biooil qualities in view to impurities, low chemical biooil stability, and phase separation. Additional technical reasons are poor plant reliabilities and availabilities.

Most FP investigations reported in the literature have been conducted with 'white' wood without bark[53]. Relatively homogeneous and reasonably clean and stable, single-phase biooils have been obtained from wood. From ash-rich lignocellulosic materials like cereal straw and other grassy biomass, we have obtained a lower biooil quality and yield with higher water content, which results in immediate or delayed phase separation into a heavier tar phase and a lighter aqueous phase [35].

In practice, two different condensates are obtained by a two-step condensation: First, a tar condensate at about 100°C with a few percent of water, which can solidify already at temperatures much above ambient. At about ambient temperature, an aqueous condensate is obtained with ca. 70 ± 15% water and various dissolved organics and has a lower heating value (LHV) of usually less than 5 MJ/kg [27]. Biooils with two phases are unsuited for higher combustion applications: Biooil contamination with pyrolysis char particles is another problem because all ash is contained in the char. Removal of the fine char particles by filtration fails by filter plugging and centrifugation by insufficient density differences.

Compared to combustion, biooil quality requirements for PEF gasification in a GSP-type gasifier are much lower. At least ca. 1 wt.% ash is even needed to generate a protective slag layer at the inner surface of the gasification chamber. Poor pyrolysis condensates with much char and ash are therefore still suited for bioslurry preparation and subsequent gasification. The pyrolysis char increases the energy content of the biooil considerably by 30% to 80%. Poor-quality lignocellulosics, e.g., ash-rich agricultural residues like cereal straw, are still available as an almost unused biocarbon resource. They can now be tapped and contribute substantially to the global biocarbon potential. The lower quality requirements connected with a change of biooil application from combustion to gasification can help to simplify the FP process.

Biomass Pyrolysis as an Independent Process

FP of biomass can also be designed as an independent process for the recovery of valuable pyrolysis products without integration into a biosynfuel production. Potential applications and recovery procedures for particular pyrolysis products are reported in the literature [54]. Commercial applications are the production of food flavorings (liquid smoke) and other fine chemicals as practiced, e.g., by Ensyn Company. A removal of a few mass percent biooil constituents is not expected to jeopardize bioslurry production for subsequent gasification. An assumed profit of only 3/kg for 3 wt.% of recovered valuable biooil constituents might cover already all technical bioslurry manufacturing costs of *ca.* 50/t (see also the 'Economic aspects' section). It is likely that such opportunities are developed and commercially applied in the future. A speculative extrapolation into an extended and established biorefinery future involves an annual biooil production globally in a gigaton range. Removal of minor constituents of a few percent in weight extends already into the ≥ 10-Mt/a production range and can create a significant contribution to the supply of organic specialty chemicals.

Reactor Types for Fp of Biomass

Various reactor types are being investigated for FP of biomass since about three decades [36,55] without a clear champion; they are depicted in Figure 6. Most types use an excess of a hot, grainy heat carrier - usually 1-mm quartz sand - heated to about 550°C, which is quickly mixed with the dry (≤ 15% water) biomass, diminuted to less than 3-mm grain size. FP takes place in about 1 s, and the pyrolysis product gas, condensable vapors, and small char particles are expelled from the heat carrier bed in about 1 s. The heat carrier grains are cooled down by $T = 10$ to 100 K to a final temperature of about 500°C and are then recycled and reheated in a closed loop. The bulk of the fine pyrolysis char particles is carried with the hot pyrolysis gases and vapors and is removed directly at the reactor exit in a hot cyclone operated at the FP reactor temperature of 500°C. A minor char fraction is retained in the heat carrier loop. With a well-designed and well-operated pyrolysis reactor, char accumulation in the heat carrier loop remains at an acceptably low level. Downstream from the cyclone, the pyrolysis gases and vapors are usually quenched to about ambient temperature by the injection of a large stream of cooled condensate through nozzles. Rapid quench cooling in a few seconds is essential to prevent significant pyrolysis vapor decomposition and maintains a high condensate yield. Quenching techniques avoid the fouling of heat exchanger walls with tar deposits. The disadvantage is that quench condensation does not allow efficient heat recovery.

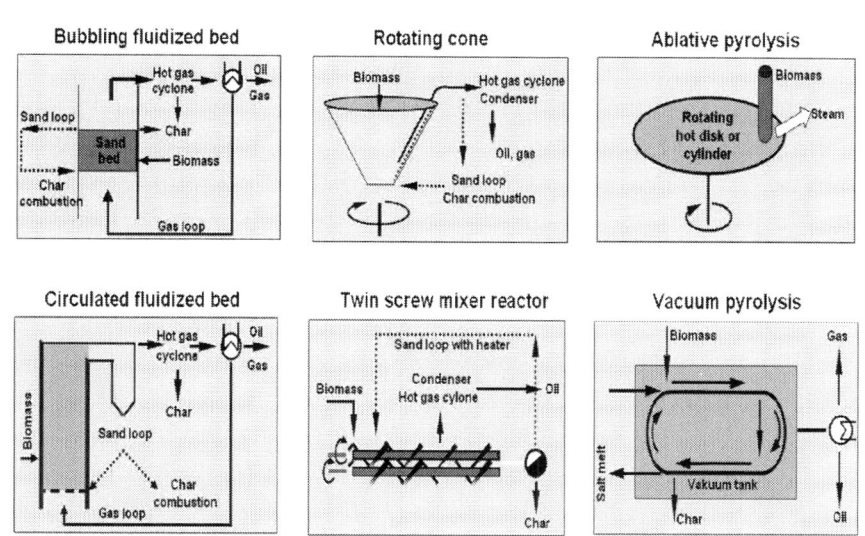

Figure 6: Reactor types used for fast pyrolysis of biomass.

The most common reactor type is a bubbling fluidized bed with *ca.* 1-mm quartz sand [36,55]. Cold pyrolysis gas downstream from the quench condenser must be recycled for bed fluidization. Pyrolysis vapor dilution with non-condensable gases increases the undesired energy loss during quench condensation and requires a larger and more expensive condensation system.

A circulating fluidized bed requires even more fluidizing gas. Ensyn Company successfully operates such 2-t/h FP reactors since many years on a commercial scale, but different to optimum syngas generation, the energy efficiency is not an important aspect for their production of fine chemicals and food flavorings.

The rotating cone reactor [56,57] and the twin-screw mixer reactor [58] use a hot heat carrier loop with a mechanically fluidized bed without an auxiliary fluidizing gas. This reduces the size of the biooil condensation system, but especially somewhat higher flow fluctuations and reduced char removal efficiency in the cyclone must be considered. Vacuum operation at *ca.* 0.1 to 0.2 bar is another more general method [59], which can be applied in all process versions to reduce the gas and vapor residence time. However, technology becomes more complex, and control of air in leakage is an additional safety aspect, which usually is prevented by a slight overpressure. Pyrovac Company, Canada has discontinued pilot plant operation because of financial problems. The state of development of ablative pyrolysis is relatively low, especially in view to scale-up [60]. The ceramic ball-heated downflow tube reactor, developed at Shandong University of Technology, China, deserves attention because of its simple design and operation [61].

The Twin-Screw Mixer Reactor

The twin-screw mixer (TSM) reactor was chosen because it was already applied on a technical scale for FP of other materials like coal, oil refinery residues, or oil shale [58]. Technology development started in the 1950s with a collaboration of Lurgi and Ruhrgas Companies for the so-called Lurgi-Ruhrgas (LR)-mixer reactor for coal pyrolysis for town gas production [62]. If the TSM reactor turns out to be suited also for FP of biomass, it is expected that the available industrial experience will contribute to reduce time and cost of further development to a commercial scale. This practical aspect does not necessarily mean that design and operating principles of the TSM are superior to the other FP reactors. Any type shown in Figure 6 is principally suited to prepare a bioslurry for PEF gasification. Also, the pyrolysis product yield structure is not expected to be much different. Final selection criteria will be based on costs, safety, reliability, and plant availability, which depend much on the FP reactor periphery.

Design characteristics of the TSM reactor are two intertwining and specially shaped screws, rotating in the same sense and cleaning each other as well as the internal reactor surfaces. Design and operating principles are outlined in Figure 7. The grainy material is transported axially and mixed radially. A suitable rotation frequency v is at a Froude number of 1. This means that the centrifugal force $2\varpi^2 \cdot m \cdot d^2$ at the outer screw radius equals the weight $m \cdot g$. This creates fluidization, which considerably eases transport and mixing. The level in the reactor increases in proportion with the throughput and is usually kept at less than half to prevent plugging.

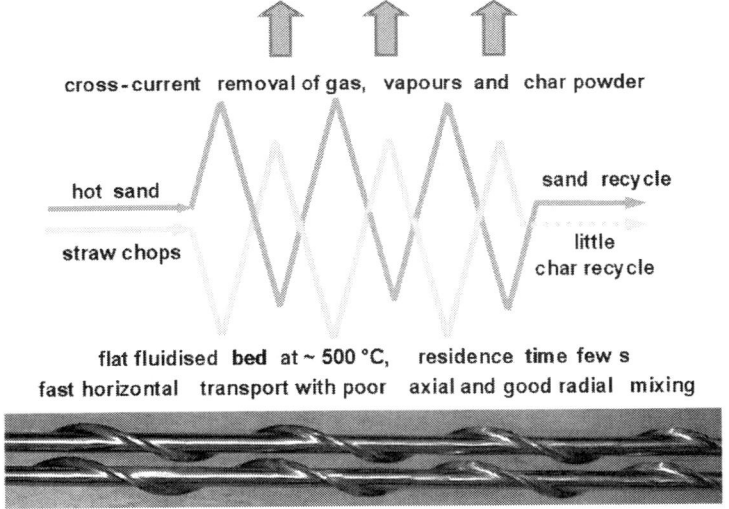

Figure 7: Principle of the twin-screw mixer reactor.

At typical residence times in the order of about 10 s, the reactor surface is too small to supply the heat for pyrolysis through the wall. A surplus of a hot, grainy heat carrier material, e.g., quartz or SiC sand, ceramic grains, or SS balls, is therefore quickly mixed with the cold diminuted biofeed. To ensure a rapid pyrolytic decomposition, the particle size of heat carrier and biofeed must be small enough to expose a sufficiently large surface for heat transfer. A desirable heat carrier/feed ratio on a volume basis is about 2; this means that the empty space between the heat carrier grains of about 40% of the total bed volume is filled with the bulky diminuted biofeed. Since the biomass volume shrinks to about half during pyrolysis, about equal bulk volumes are a reasonable maximum at start. With a bulk density of 4,800 kg/m^3 for steel balls and 100 kg/m^3 for un-pyrolyzed straw chops, about 50 kg of steel balls will be circulated per kilogram of biomass. This is the design ratio in our FP-process development unit (PDU). All pyrolysis gases, vapors, and fine char particles are expelled in a cross-flow direction from the shallow reaction bed. Rapid removal and quench condensation of the pyrolysis vapors is essential to prevent thermal vapor decomposition at the surfaces of the hot heat carrier grains and maintains high condensate yields.

Pyrolysis Facilities at KIT

Lab-Scale Fluidized Bed

For quick screening tests of the FP behavior of various biomaterials, a lab-scale device with a bubbling fluidized sand bed for a maximum of 0.3 kg/h of throughput has been built (Figure 8). The reactor is 4 cm in diameter and is filled 12 cm high with *ca.* 0.2-kg, 0.2- to 0.3-mm-diameter quartz sand and fluidized with 1 m^3(standard temperature and pressure (STP))/h preheated nitrogen. A pre-weight amount of *ca.* 0.5 kg of diminuted biomass is constantly fed into the fluidized bed with a screw feeder together with a slight nitrogen stream to prevent backflow of pyrolysis gas. The pyrolysis reactor and the subsequent char cyclone are mounted in an electrically heated oven. Product recovery is conventional via a hot cyclone and a two-stage condenser with an electrostatic precipitator. At the end, the mass of char, condensate, and gas is determined and analyzed.

Figure 8: Laboratory-scale FP device.

Process Development Unit

In 2002 to 2003, a PDU with a TSM reactor for a throughput of 10 to 20 kg/h of biomass has been designed and built at the KIT to test the suitability of the twin-screw reactor type for FP of biomass[26]. A simplified flow sheet is shown in Figure 9. The major plant sections are briefly described.

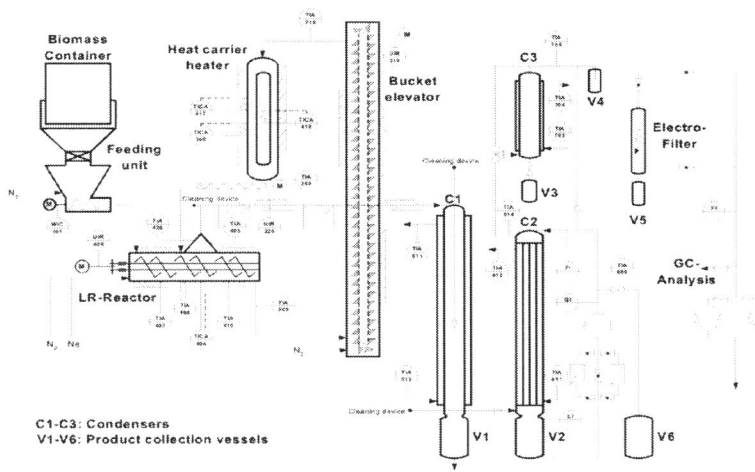

Figure 9. Flow sheet of the FP PDU.

- *Hot heat carrier loop.* A grainy heat carrier circulates at a temperature of about 500°C in a closed, gastight loop with a single exit for all pyrolysis products. Various heat carriers either 1-mm sand or SiC grains or 1.5-mm SS balls are lifted vertically 3 m with a conventional bucket elevator made from SS. The heat carrier material is reheated by $\Delta T = 10°C$ to 100°C during gravity flow through a 1-m-high, coaxial twin cylinder with a diameter of 0.15 m and a 1-cm-wide annular gap, heated electrically from both sides via a 1-m² surface. A volume-calibrated, controlled screw feeder transports a constant heat carrier stream into the pyrolysis reactor, at a maximum of either 0.4-t/h, 1-mm quartz (bulk density 1,500 kg/m³) or SiC sand or 1.5- to 2-mm SS balls up to 1.5 t/h (bulk density 4,800 kg/m³). A second screw feeder controls the biomass feed rate of 10 to 20 kg/h. Main construction material in the hot loop section is SS, which turned out to be suitable.

- *TSM reactor.* The active length of the twin-screw reactor is 1 m; the inner and outer screw diameters are 2 and 4 cm, respectively; and the pitch is 0.2 m (see Figure 8). A typical rotation frequency is *ca.* 3 Hz (Froude number almost 1). The heat carrier mean residence time in the reactor of *ca.* 10 s is almost independent from the heat carrier flow as long as the heat carrier level is below about half. Kinetic measurements have shown that this time is sufficient for FP. Figure 10 shows that the pyrolysis rate for < 2-mm wood particles are faster, especially for cereal straw which has a wall thickness of only 0.3 mm (0.15 mm characteristic length). From the bulk volume flow rate at typical operating conditions, it has been estimated that the reactor volume is usually filled up to only less than half, a sufficiently low level to prevent plugging.

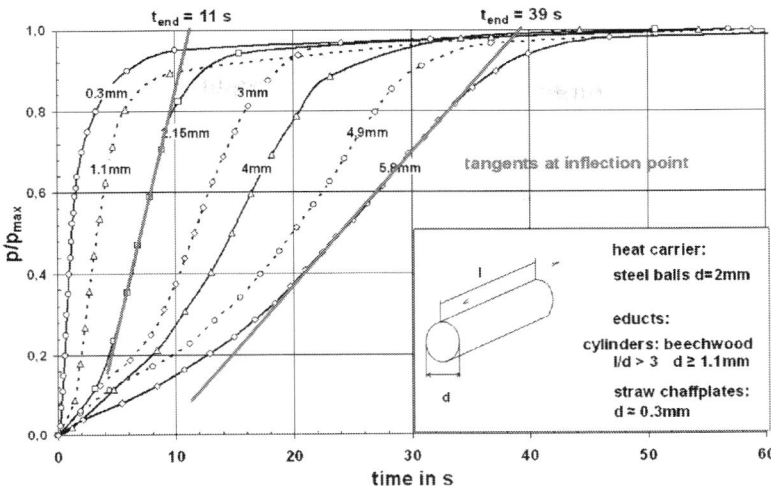

Figure 10: Pyrolysis kinetics of wood and straw.

- *Product recovery system.* The normal product recovery system consists of a hot cyclone operated at a reactor temperature of 500°C to remove the bulk of the entrained char particles. This is complemented by a subsequent quench condenser for flash condensation of tars and reaction water by the recycle and injection of a cooled quench condensate. In the KIT PDU, this system is frequently modified and tested in an iterative process to find the best way for a reliable recovery operation.

Trouble with solid deposits can arise if sticky tars condense at the walls and collect char powder from the gas stream. At higher temperatures, the soft deposits decompose gradually to a hard, black, and highly porous material. Automatic or occasional mechanical removal of potential deposits is advisable at few critical sites to maintain a reliable continuous operation without interruption. The flow sheet in Figure 9 shows the actual test version for product recovery: after quick cooling to *ca.* 100°C in the presence of char, char crumbs are removed with condensed tar soaked and eventually solidified in the pore system. The more or less solidified tar in the pores deactivates the char and prevents self-ignition and char dust inhalation during handling.

Lurgi's Mini-LR Plant

In addition to the operation of the KIT PDU, we have performed an experimental campaign at the 3- to 5-kg/h Mini-LR plant of Lurgi Company in Frankfurt. The main difference of the two facilities shown in the photos of Figure 11 is the design of the heat carrier loop, as outlined in Figure 12. Heat carrier in the Mini-LR plant is 1-mm quartz sand. It is lifted pneumatically with hot flue gas from pyrolysis gas combustion with air and simultaneously reheated to a maximum temperature of 600°C in direct contact with excellent heat transfer.

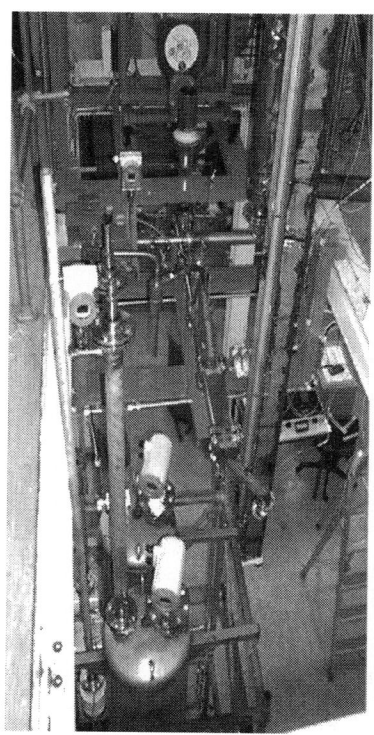 

Figure 11: Photos of the FP PDU at KIT (left) and of Lurgi's Mini-LR plant (right).

PDU plant (15 kg/h), KIT version

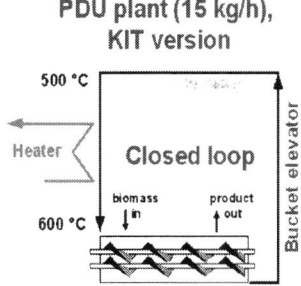

Bucket elevator, indirect heating:

+ No restriction for heat carriers regarding particle size and density.

- Low heat transfer through heat exchanger walls (surface↑, costs↑).

Pilot plant at KIT (500 kg/h), Lurgi version

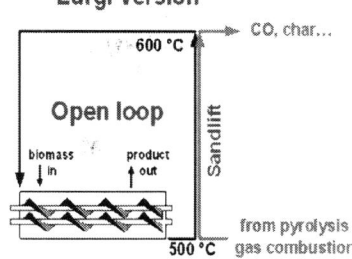

Pneumatic sandlift, direct heating:

+ Good heat exchange conditions and simultaneous transport.

- Expected higher attrition.

- Formation of pollutants (CO).

Figure 12: Essential differences of the KIT and Lurgi FP heat carrier loop concepts.

Because the flue gas has been in contact with pyrolysis residues in the heat carrier sand and is afterwards released into the atmosphere, the system is open to the environment and needs careful gas cleaning especially after contact with the char-contaminated heat carrier grains. To prevent the intrusion of the slightly pressurized lift gas into the pyrolysis reactor, it must be separated above and below by the flow resistance of a longer, sand-filled, pipe section of several meters in length. This increases the height and cost of the expensive hot loop section. Sand particle attrition must also be considered because of the high velocities of almost 20 m/s in the lift pipe. Successful industrial experience is claimed for this version.

FP Pilot Facility at Karlsruhe

Mid-2005, after experimental confirmation of the principal suitability of the TSM reactor for biomass FP in the small KIT and Lurgi FP facilities and after four successful bioslurry gasification campaigns in the 3- to 5-MW(th), GSP-type, PEF pilot gasifier at a 26-bar pressure with up to 0.6 t/h (3 MW(th)) of bioslurry throughput (see the 'Bioslurry gasification' section), it has been decided to extend also the small-scale FP investigations to the pilot plant scale to determine design data for a FP demonstration plant.

A 0.5-t/h FP pilot plant (2 MW(th)) based on Lurgi's industrial experience [63] with sand as heat carrier and a pneumatic lift in an 'open loop' version has been built up at KIT. A simplified flow sheet is shown in Figure 13. Figure 14 shows a photo, and in the study of Dahmen [64], a brief description is given. The plant is in operation since 2010 in test campaigns of typically 1 week in duration using straw as the feed material.

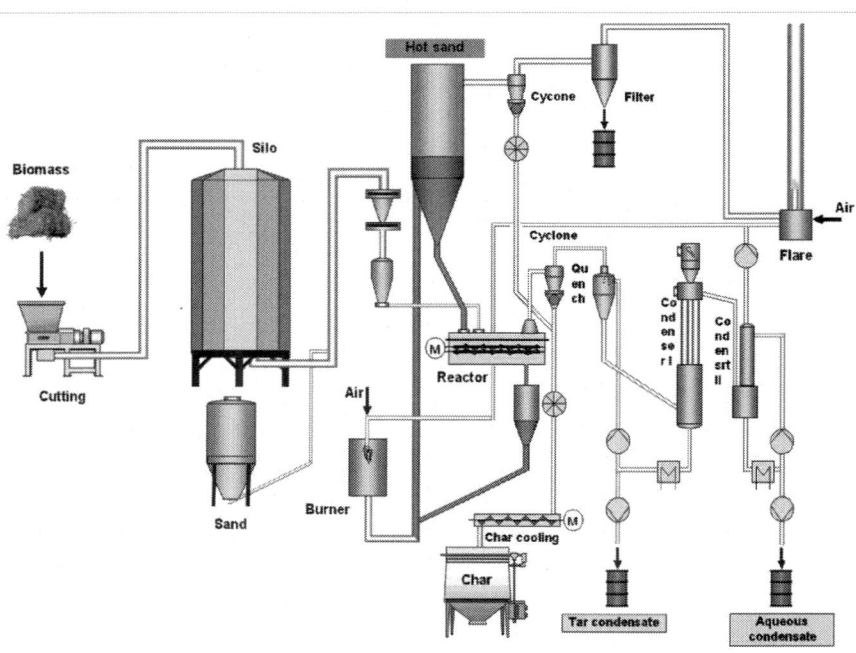

Figure 13: Simplified flow sheet of the FP pilot plant at KIT.

Figure 14: Photos of the 0.5-t/h FP plant at KIT.

Experimental Results and Operating Experience

Typical Operation Conditions

Meanwhile, the accumulated operating experience for FP of biomass in the PDU amounts to more than 2,000 h of operation with more than 100 individual runs and more than two dozens of different biomass types, e.g., hardwood, softwood, wheat, maize, straw, rice straw, hay, miscanthus, bran, different oil palm residues, sugar cane bagasse, etc. A typical run starts with the preheated facility and the heat carrier circulating in the loop at the correct operating temperature and circulation rate. At a feed rate of 10 to 20 kg/h of dry diminuted biomass, it takes several hours until a carefully pre-weight total amount of 40 to 80 kg of biomass is fed at a constant rate into the pyrolysis reactor. Several hours of steady-state operation turned out to be sufficient to get a reasonably accurate mass and energy balance for the solid, liquid, and gaseous products, whose percentages and properties are needed for the subsequent slurry preparation and gasification steps.

Char, Condensate, and Gas Yields

A typical example of yield results for a FP campaign with a total of 19 runs for four different feedstocks is summarized in Figure 15[35]. The bars represent the average yields of three to seven identical runs for each feedstock. The same type of results is shown in Figure 16 for an experimental campaign in the Mini-LR plant of Lurgi Company, Frankfurt, performed in collaboration with KIT [26]. The results are consistent within the error range. The mass yields of the liquid condensates from wood pyrolysis are three to four times higher than those of char and are more than sufficient to produce a free-flowing bioslurry (see the 'Bioslurry preparation' section). The yield of pyrolysis liquids from straw is much lower and only about twice the mass of char. At the expense of lower condensate yields, pyrolysis gas and char as well as the reaction water yields for straw are about 1.5 times higher than those for wood. The amount of reaction water plus moisture in the condensates has been determined by Karl Fischer titration.

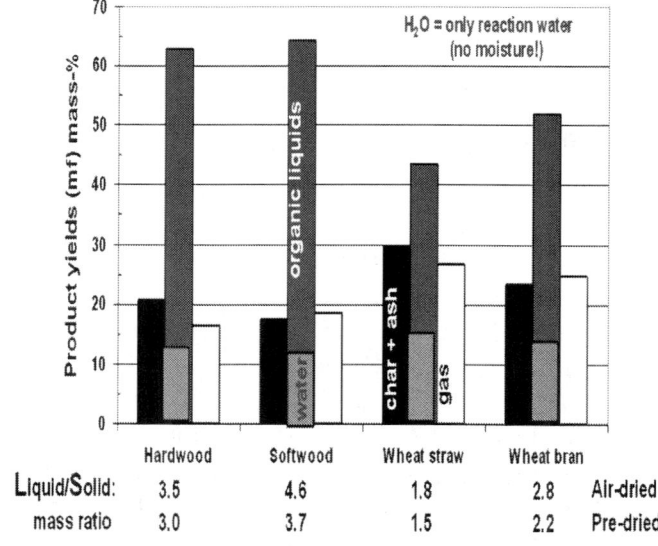

Figure 15: FP product yields for different types of biomass (KIT PDU) [35].

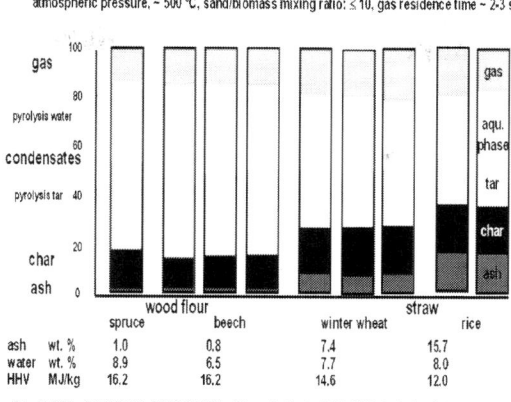

atmospheric pressure, ~ 500 °C, sand/biomass mixing ratio: ≤ 10, gas residence time ~ 2-3 s

		spruce	beech	winter wheat	rice
ash	wt. %	1.0	0.8	7.4	15.7
water	wt. %	8.9	6.5	7.7	8.0
HHV	MJ/kg	16.2	16.2	14.6	12.0

E. Henrich, E. Dinjus, R. Stahl, F. Weirich (FZK, ITC CPV); H. Weiss, U. Zentner (Lurgi-Lentjes); D. Meier (BFH-Hamburg), to be published

Figure 16: FP product yields for different types of biomass (Lurgi's Mini-LR plant).

The stability of pyrolysis condensates towards phase separation into a heavy tar phase and a lighter aqueous phase decreases with increasing water content. Above 30 to 35 wt.% water, phase separation occurs almost immediately after condensation; we could never obtain a stable biooil from air-dry cereal straw. Even for some initially homogeneous biooil phases from wood with around 25 wt.% water, we have observed a delayed phase separation after several weeks or months.

Pyrolysis Gas Composition

The typical composition of FP gases is shown in Figure 17. Main constituents are CO_2 and CO. The minor constituents H_2, CH_4, and the gaseous hydrocarbons (C_2-C_5 alkanes and alkenes) contribute about half to the heating value. Vapors of very volatile CHO constituents like formaldehyde, acetaldehyde, acetone, methanol, glyoxal, methyl, ethyl esters of formic and acetic acids, etc. can escape with the gases, but have not been analyzed in detail so far. They are suspected to be responsible, at least partly, for the typical mass balance deficit of several percent observed when all measured constituents are summarized. Thus, a simple increase of the final condensation temperature can increase the energy content in the pyrolysis gas. Thus, an in-line control of the condensation temperature can be used to adjust the energy content in the pyrolysis gas exactly to the demand for a self-sustained process.

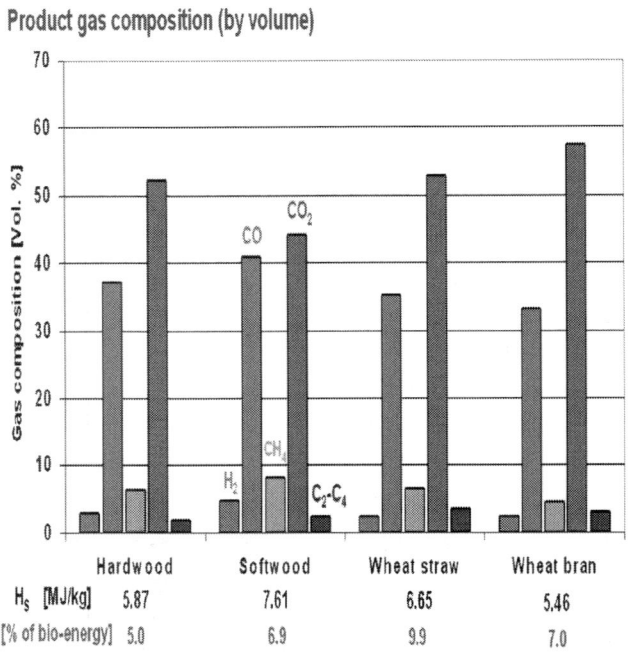

Figure 17: Typical composition of FP gas (complement to Figure 16).

Heat Required for FP

Much effort has been devoted determining the specific heat required for FP. The reaction enthalpy ($\Delta_r H$) of an exo- or endothermal pyrolysis reaction must be subtracted from the sensible heat required to heat the pyrolysis products from ambient to 500°C. The higher the yield of the combustion products CO_2, H_2O, or the related char product in reaction 1 below, the more exothermal the pyrolysis reaction becomes. This can be illustrated with the idealized pyrolysis reactions in Table4, using the experimental heat of combustion ($\Delta_c H$) of lignocellulose with the simplified formula $C_3H_4O_2$ ($\Delta_c H$(= HHV) = -1,402 kJ/mol (according to the Channiwala relation [63]), $\Delta_c H = 1,460$ kJ/mol (experimental)) and the tabulated heats of combustion for the products.

Table 4: Reaction enthalpy of idealized pyrolysis reactions

Pyrolysis	Reaction			$\Delta_r H$ (kJ/mol)	
Exothermal pyrolysis					
Reaction 1	$C_3H_4O_2 \rightarrow 3C + 2H_2O$			-278	
Reaction 2	$C_3H_4O_2 \rightarrow C + CH_4 + CO_2$			-196	
Endothermal pyrolysis					
Reaction 3	$C_3H_4O_2 \rightarrow C + 2CH_2O$			76	
$\Delta_c H$ (HHV; kJ/mol)					
	C	CH_4	CH_2O	H_2O	CO_2
Known from tables	-394	-890	-571	0	0
Channiwala estimate	-419	-895	-491	-72	-89

A self-sustained slow pyrolysis of completely dry, preheated wood in a rotary kiln was practiced commercially at Ford Motor Company until the 1930s [7,65]. For FP with much lower char, CO_2, and water yields, a thermoneutral or even endothermal pyrolysis reaction is more likely and depends on the product composition. The Channiwala relation overestimates the HHV of CO_2, H_2O, and char. In reality, the HHV of products are higher (less negative) and push $\Delta_r H$ towards a more endothermal value.

In the PDU, the overall heat consumption has been measured experimentally for various biomaterials. The amount of heat consumed for pyrolysis corresponds to the heat removed from the heat carrier, which is the known product of the heat carrier flow rate m· (in kilograms per second), the specific heat at reaction temperature c_p (ca. 0.7 kJ/kg K for SS at 500°C), and the temperature difference ΔT between the reactor inlet and exit. In a small facility, the simultaneous heat in- or outflow through the reactor insulation must also be considered.

The heat required for the pyrolysis of 1 kg of biomass is obtained by division with the biomass feed rate in kilograms per second. Figure 18 shows an example for the temperature drop ΔT in the reactor during a 4-h, stationary-state operation (60 to 300 min) in a typical run with a constant of 1.14-t/h SS ball heat carrier circulation at a 9.5-kg/h hardwood feed rate.

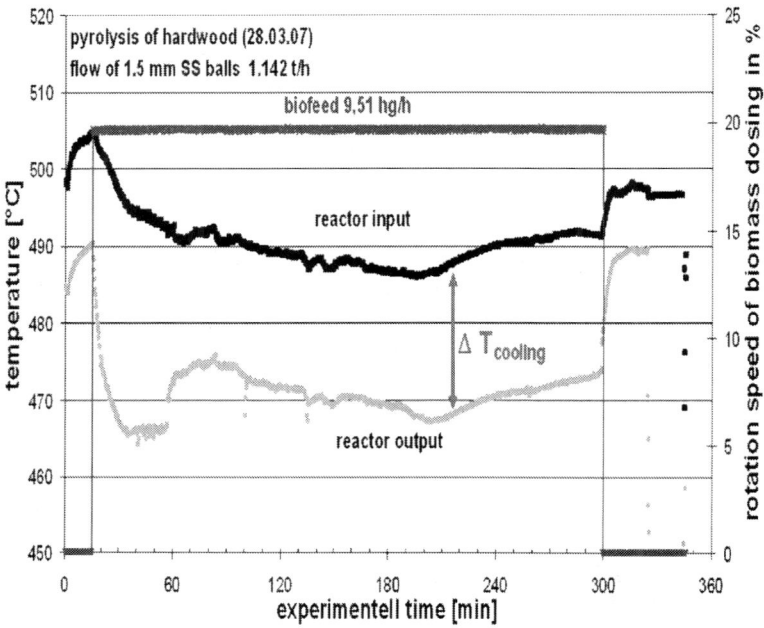

Figure 18: In- and outlet temperatures of SS ball heat carrier.

A mean heat consumption of 1.3 ± 0.4 MJ/kg for FP has been measured for dry lignocellulosics [35]. Daugaard [66] has reported a range of 0.8 to 1.6 MJ/kg for FP for various lignocellulosics. This corresponds to about 7 ± 2% of the initial bioenergy. For moist materials, the value is somewhat higher since water needs 3.4 MJ/kg for heat-up from 20°C to 500°C. Antal and Gronli [67] have reported an about linear increase of the liberated reaction heat with increasing char yield with thermoneutrality in the range of about 20% char yield. Together with some thermal insulation losses, a consumption of *ca.* 10% of the bioenergy is therefore expected for FP in practice. The pyrolysis gases contain 6% to 10% of the initial bioenergy without volatile oxygenates, and their combustion should supply sufficient energy for FP, at least at a somewhat higher final condensation temperature. Thus, all char and condensates remain available for bioslurry preparation, and there is no process waste except the flue gas from pyrolysis gas combustion. At the end, the pyrolysis gas is washed with the aqueous pyrolysis condensate and will be relatively clean.

Quartz Sand and SS Balls as Heat Carriers

We have experimentally compared 1-mm quartz and SiC sand and 1.5-mm SS balls as heat carrier materials. The essential experience is that the SS balls are superior: The throughput of the PDU could be increased by at least 50%, and the availability of the facility and the reliability of operation could be improved considerably. The small amount of attrited fine sand, especially from SiC particles, caused some erosion in downstream pumps. The SS balls did not show attrition; after 1,000 h of operation, the mass of several hundreds of clean steel balls did not change within error, a hint to low wear and tear.

Modeling of Lignocellulose FP

The optimum FP temperature of *ca.* 500°C for lignocellulose is in the range where C-C bonds can form and break simultaneously. The decomposition of the biopolymer structures by FP creates a complex multitude of hundreds of different solid, liquid, and gaseous carbon species. We do not rely on any speculations concerning macro-kinetic reaction mechanisms to predict the lumped yields of char, organic condensate, reaction water, and gases for the various lignocellulosic bio-feedstocks in this complex system. Based on literature data and our own experiments, we have decided for a rather primitive and oversimplified yield prediction model for FP of lignocellulose at atmospheric pressure and at 500°C in a well-designed FP reactor. Lignocellulosic materials are divided into only two groups according to their ash content: < 2 wt.% (e.g., wood) and > 2 wt.% (e.g., straw). The lumped product yields for the water and ash-free lignocellulosic CHO fraction are given in Table 5.

Table 5: FP yield prediction for lignocellulose given on a water- and ash-free basis

Ash content	Solid	Liquids		Gas	Sum (%)
	Char (wt.%)	Tar (wt.%)	Reaction water (wt.%)	Gas (wt.%)	
Low, < 2% (e.g., wood)	16	56	12	16	100
High, > 2% (e.g., straw)	24	34	18	24	100

For a moist lignocellulose with ash, the ash must be added to the char, and the moisture, to the reaction water or the condensate. For the yield percentages related to the real material, the corrected numbers must be normalized to 100%. For highly ash-containing lignocellulose like straw, the char, gas, and reaction water yields increase by a factor of roughly 1.5, at the expense of the much lower tar yield. This can be explained by catalytic tar vapor decomposition at the ash and char surfaces. The relatively large yield fluctuations observed in practice indicate that the controlling feed properties and operating conditions are not yet completely understood. Table 5 is helpful to get a rough first estimate without much effort; experimental confirmation must follow.

Scale-Up of the Tsm Reactor

Eight LR mixer reactors have been built by Lurgi Company, Frankfurt up to a 1-m screw diameter and 600-m^3/h heat carrier circulation [58]. Reliable scale-up rules based on similarity criteria [68] have been developed by Peters [62]. The volumetric feed rate V· and reactor volume V should scale with the outer screw diameter d according to the following equation: $V_1/V_2 = (d_1/d_2)^{2.5}$. An extreme and therefore less reliable extrapolation from a small 20-kg/h design in the PDU to 20 t/h in a large commercial plant corresponds to a 1,000-fold feed rate increase and an about 16-fold increase of a 0.04-m screw diameter to ca. 0.64 m and a 16--√-fold screw length increase from 1 to ca. 5 m. A screw pitch of 2.5 m corresponds to a length/diameter ratio of 4. At a Froude number of 1, the rotation frequency in the 20-t/h plant is $n = g/(2\varpi^2 d)^{1/2} = 1.15$ Hz. At a 12-s mean residence time and a 1,000-t/h SS ball circulation (ca. 210 m^3/h or ca. 0.06 m^3/s), the ball inventory is about 0.36 m³ or *ca.* 30% of the reactor volume.

Per kilogram of lignocellulose, 50 kg of SS balls are circulated, which requires 1.3 ± 0.4 MJ of heat. With the specific heat for SS of 0.7 kJ/kg·K at 500°C, a heat-up of $\Delta T = 37$ K is needed for 1.3 MJ. In view to some thermal loss, a temperature difference of at least 40 K seems reasonable; to cover extreme situations, a design value of about $\Delta T = 50$ K is advisable.

A twin-screw reactor with corrosion-resistant Incoloy 800 plated screws and a refractory liner at the reactor wall represents ≤ 10% of the total capital investment (TCI) for a FP plant. The remaining design and operating problems are expected to be proportional to the remaining capital expenditure (capex) of 90%, and potential problems arising in the reactor periphery should therefore not be underestimated. This statement is amplified by the fact that experience from large technical FP plants with the desired capacity does not exist; their state of development is the lowest in the bioliq process chain.

Bioslurry Preparation

Gasifier Feed Preparation Options

The aim of the bioslurry gasification concept is the preparation of a convenient feed for a large PEF gasifier. For this purpose, it is unimportant if the FP plants are colocated at the gasifier site or distributed in the region. The latter option is attractive since the dense pyrolysis slurries or pastes are suited for easy handling, compact storage, and cheap transport, which favor the erection of a large and more economic central BTL plant. Optimum slurry properties are not necessarily the same for handling, storage, transport, or gasifier feeding. A stepwise preparation of the final feed form is therefore a good design guide. Final slurry preparation steps which are reasonable only at the gasifier site are heating for viscosity reduction using the abundant waste heat in the BTL plant and by feed temperature rise by about 10 K or more during efficient colloid mixing prior to feeding to ensure homogeneity without leaving sufficient time for sedimentation. Separate feeding of solid, liquid, or gaseous FP products to a gasifier is less favorable.

Pyrolysis Gas Feed

Compression of dirty pyrolysis gases with much tar is difficult to realize for PEF gasification. If pyrolysis gases are fed to a large gasifier, the pyrolysis plants must be colocated. All biomass must therefore be delivered over larger distances directly to the central BTL plant. Increasing transport costs set an economic limit to the BTL plant size. A reasonable maximum biomass delivery radius for residual straw and forest residues in central Europe of about 70 km corresponds to *ca.* 1.4 Gt/a with a very small biosynfuel production capacity of only 0.2 Mt/a or 2% of a modern oil refinery. The same amount of biomass can be harvested in an energy plantation of only 30 km in diameter at an annual harvest of 20 t/ha. An advantage of colocation is that sufficient low-temperature waste heat is available in a BTL plant for drying moist feedstock like fresh wood with *ca.* 50 wt.% water, but storage of moist biomass above about 15 wt.% water is not possible without some biological degradation.

Pulverized Pyrolysis Char Feed

Firing of pulverized, 50-μm coal is standard in large, coal-fired power stations. Coal particles are not porous, and large volumes of combustion air are available for pneumatic transport at atmospheric pressure. Different from that, pyrolysis or torrefaction char powders are highly porous and have a low pneumatic transport density of less than 300 kg/m^3. Char powder transfer from a pressurized fluidized bed therefore requires large volumes of recycled inert syngas. This increases the raw syngas flow and the oxygen consumption, thus reducing the efficiency of syngas generation. The gasification agent oxygen cannot be used for char powder transport for safety reasons. Handling of large amounts of reactive pyrolysis char powders requires careful technical control of safety hazards like potential self-ignition, dust explosions, or inhalation of fine, filter-penetrating char dust. Intermediate pelletization improves safety and handling during storage and transportation.

Biooil Feed

FP biooils from wood contain up to two-thirds of the initial bioenergy, biooils from ash-rich lignocellulosic materials like straw contain only about half. Straw biooils are less stable and prone to immediate or delayed phase separation. This is a general characteristic of poor biooils with more than 30 to 35 wt.% water.

The heating value of the lighter, aqueous-phase condensates is too low for safe gasification. To increase syngas generation efficiency, biooil gasification must be supported by pyrolysis char addition. In a bioslurry, both products are combined in a single, dense stream which can contain up to *ca.* 90% of the bioenergy and simultaneously permits safe and easy handling of the reactive pyrolysis char powder.

Bioslurries, Pastes, and Char/Tar Crumbs

In the bioslurry concept, the pyrolysis gas supplies the thermal energy for the FP process. The gas carries small amounts of volatile pyrolysis vapors whose removal efficiency can be controlled by the final condensation temperature. Thus, the combustion energy can be exactly adjusted to the actual need. All non-gaseous products such as char, tar, and aqueous condensates are mixed together to a single, free-flowing slurry, pasty mud, or sludge stream. A small volume of condensate is completely soaked up, and char crumbs are formed. Any other waste liquids, slurries, pastes, or sludges with some heating value can also be used to slurry a char.

Bioslurries

Bioslurries are free-flowing mixtures of pyrolysis condensates and pyrolysis char powders below 100 μm in size. The char powder content should be as high as reasonably possible, normally at almost sedimentation density. Slurries with a particle volume fraction up to *ca.* 50% are still free-flowing and pumpable. This volume percentage is somewhat below the *ca.* 60 vol.% of solids in a bed of regular crystals or the 75 vol.% for a dense packing of spheres. Little volume expansion is sufficient to allow free particle movement similar to a fluidized bed. Pyrolysis chars have a high porosity between about 50% and 80%. The pore system first soaks up much liquid like a sponge until a sufficient volume remains as a 'lubricant' outside the particles (as visualized in Figure 19). As a rule of thumb, a liquid/solid (L/S) volume ratio of 1 corresponds to a biooil/char L/S weight ratio of about 2 to 3 in the slurry. This weight ratio is usually sufficient to prepare free-flowing slurries without much effort, at least after warming and colloid mixing. Pumpable slurries of nonporous, pulverized coal and water have been considered as a type of raw oil substitute during the oil crisis, e.g., for SNG production and have been prepared down to a L/S weight ratio of 0.4.

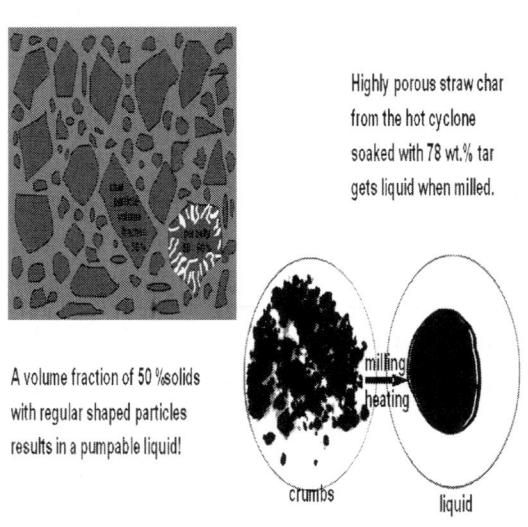

Figure 19: Principles of slurry preparation.

A first crude estimate of the maximum char loading can be obtained from the bulk density of a slightly compressed pyrolysis char powder column, 300 to 400 kg/m³ corresponds to an L/S mass ratio of 2 to 3 at typical bioslurry densities around 1,200 kg/m³. During FP of wood without bark, typical L/S ratios of 3 to 5 have been attained; this is more than sufficient for slurry preparation. During FP of cereal straw with 5% to 10% ash, L/S ratios of about 2 have been measured, which is just at the acceptable limits. An L/S ratio of *ca.* 1.5 found for rice straw with 15% to 20% ash requires some additional efforts. The success depends much on the special properties of char and biooil; special bioslurry preparations with an L/S ratio as low as 1.6 have already been realized on the 1-t scale. Separate handling of some surplus char is another practicable option.

Small char particles with a broad size spectrum from 100 µm down to *ca*. 1 µm allow higher loadings. The same is true for a bimodal size distribution, where the *ca*. 10 times smaller particles can pass into the interstices of the larger grains. Using steel balls with diameters of a few millimeters as heat carrier, the TSM acts like a ball mill and produces a fine powder from the brittle pyrolysis char. A substantial part of the pyrolysis chars are larger aggregates of smaller porous char particles. Deagglomeration in a colloid mixer liberates some interstitial liquid for use as a lubricant. A combination of char diminution to less than 100 µm, deagglomeration, and homogenization in a colloid mixer combined with stepwise warming and further in-line preheating under pressure immediately before gasification are essential ingredients for a successful final bioslurry preparation.

Sedimentation

Slurries with small char particles in a size range of a few tens of micrometers and below, mixed into a relatively viscous, single-phase biooil, remain homogeneous for weeks or months. Larger char particles and char crumbs suspended especially in aqueous condensates with a low, water-like viscosity tend to undergo sedimentation in the course of an hour or a day. Sedimentation can be prevented by the addition of a gelling agent, e.g., about 3% flour (starch) [69]. A simpler measure for unstable slurries is remixing immediately before gasifier feeding and immediate flushing of feed pipes after a shutdown or a feed stop.

Pastes

Pastes, muds, or sludges are generated when the char content exceeds the sedimentation density at a L/S weight ratio below about 2, but this is not a sharp limit. The rheological properties of these concentrated biooil/char slurries, pastes, muds, or sludges are not known well enough at present to justify much discussion.

Pastes are not free-flowing and therefore do not run out in case of a vessel leak and are not soaked into the soil in case of an accidental spill. In case of frost, especially during transport, a paste freezes more slowly and with less dehomogenization. These are desirable properties for storage and transport. Transfer with pumps must be replaced by a more rapid gravity discharge from silo transport containers into a vessel below. Transfer pumps might also turn out to be too slow and expensive even for simple discharge of large slurry volumes.

Char/Tar Crumbs

Char/tar crumbs are a special handling form, which is being investigated with the aim to simplify the whole FP product recovery procedure. Char/tar crumbs are obtained by a partial quench of the FP product stream to about 100°C immediately downstream from the pyrolysis reactor. For that purpose, a controlled fine spray of aqueous FP condensate is dispersed for fast evaporation cooling in the presence of all char. The condensed high-boiling tar constituents are soaked into the char pore system and eventually solidified there. This reduces the pore volume and the amount of liquid required for slurry preparation later on. After the removal of the char crumbs from the gas stream, an aqueous pyrolysis condensate is recovered by conventional means. A clean part of this aqueous condensate is recycled for spray quenching in the first stage, and the remainder is used for slurry preparation. No hot cyclone is needed for this process version.

Lab- and Pilot-Scale Bioslurry Production

Various bioslurry production and characterization procedures are being investigated in the laboratory and have been used on pilot plant scale. This section confines to some essential aspects; more details have been reported in the literature [37,38].

Lab-Scale Investigations

Essential principles are depicted in Figure 19. The importance of (1) the minimum L/S weight ratio of about 2, (2) char particle deagglomeration in a colloid mixer by high shear stress in the order of $\geq 10^{-4}$/s, and (3) heating for viscosity reduction have already been mentioned. Many viscosity measurements have shown that a temperature increase of 50 K decreases the slurry viscosity by about an order of magnitude as shown in Figure 20. Heating a bioslurry under pressure by at least 100°C can reduce a high, room-temperature viscosity of 30 Pa s down to 0.3 Pa s, a sufficiently low viscosity for efficient pneumatic atomization with oxygen in the special gasifier feed nozzles.

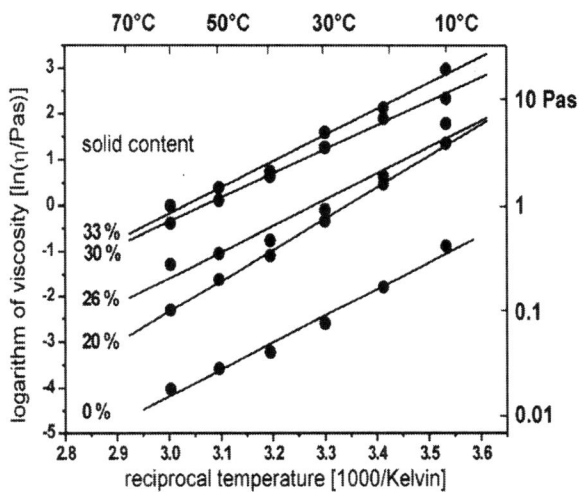

Figure 20: Temperature dependence of viscosity for various bioslurries.

Some observations are briefly summarized: Bioslurries have the same unpleasant smoky and acrid smell as the biooils and must be handled in closed systems. During phase separation of poor-quality biooils, a heavy, hydrophobic tar phase settles at the bottom and collects most of the suspended char particles. In rare cases, a small volume of a third phase of light hydrocarbons (e.g., turpentine-like) also accumulates simultaneously as a thin layer at the surface of the aqueous phase [70,71].

An important progress on the way to higher char loadings and more homogeneous and stable bioslurries was the use of a colloid mixer. The machine has been designed by MAT Company, Immenstadt, Germany, for the robust production of concentrated cement lime in the building and construction industry. The slurry, 2 to 5 kg in the lab-scale colloid mixer, is contained in a continuously stirred tank reactor and passes several times through a section with rapidly rotating, perforated paddles with narrow, ca. 1-mm gaps to the wall and a high shear field. The robust construction and operation and the rough power for the high shear rate exerted heat the slurry by some kelvin or more than 10 K, increasing with the mixing time and speed.

Bioslurry Production on the 1-T/H Scale

For the bioslurry gasification campaigns in the 0.3- to 1-t/h PEF pilot gasifier at Freiberg, a total of about 40 t of pumpable bioslurries have been prepared from different biomass and different pyrolysis conditions. From the few and sparsely operated FP pilot facilities, such ton amounts are hardly available. In most cases, we have therefore used the comparable products from conventional beech wood pyrolysis [72] for charcoal production with the Degussa process (proFagus Company, Bodenfelde, Germany): A grainy charcoal fraction was milled in a ball mill to various sizes from 100 down to 10 μm. An organic condensate with a composition similar to a single-phase FP biooil from wood and an aqueous-phase condensate with 70% water with dissolved organics, mainly acetic acid, resemble the aqueous phase of phase-separated, low-quality biooils [24]. A real FP biooil from hardwood has been obtained from Dynamotive Company (Vancouver, Canada). At KIT, several tons of char and biooil have also been produced from cereal straw pellets in an auger reactor which was partly heated electrically from outside and partly by hot steel balls of 15 mm in diameter [73]. This process with a few hundreds of seconds of residence time is termed intermediate pyrolysis mode and achieves only slightly lower liquid yields compared to FP.

For the very first slurry production for pilot-scale gasification experiments, several tons of biooil were continuously circulated at ambient temperature with a screw pump via a 1-m³ PE mixer vessel equipped with a simple propeller agitator and a 2-kW electromotor. Charcoal powder was continuously added near the stirrer shaft until the desired slurry loading was attained. In the next bioslurry gasification campaign, we have used a commercial 0.25-m³ batch colloid mixer from MAT Company for a slurry production rate of 0.5 t/h. In the following campaigns, we have used a continuous colloid mixer for a maximum of *ca.* 1-t/h bioslurry production from the same company. The whole mixer station is shown in Figure 21 and will become part of the bioliq pilot facilities.

Fig■re 21: Continuous 1-t/h bioslurry preparation with colloid mixer.

Process Integration

A arge-scale BTL plant for 1-Mt/a biosynfuel production in 8,000 h/a needs about 600 t/h or 500 m³/h (0.14 m³/s) of bioslurry. A delivery of bioslurries or pastes with whole trains (24 wagons with a total of 48 silo containers with a 20-t capacity) requires about 1,000 t. Train delivery every 1.5 h is a tolerable traffic density. Secure unloading of a 20-t silo container every 2 min can be managed by gravity unloading with at least two unloading stations. Sugar beet discharge from the delivery trucks in a modern sugar mill is a comparable process.

The gasifier feed can be prepared batchwise in cyclic operation in several huge bioslurry mixing vessels. A carefully selected slurry menu for ca. 1-h gasifier operation is prepared by mixing and warming up ca. 30 preselected transport silos. The adjusted, warm, and viscous bioslurry is transferred with screw or plunger pumps to a colloidal mixer for efficient final homogenization. Further slurry transfer to the gasifier is then performed under pressure via a pressurized slurry heater. The high gasifier feed rate of 0.14 m³/s into the gasifier burner corresponds to a slurry injection area of 0.07 m² at a 2-m/s injection velocity. The design of the large multi-burner head in the gasifier is a challenge. A dangerous, accidental, sudden, and total feed interruption is unlikely with so many feed points (see safety consideration in the 'Further operation and development aspects' section).

Co-firing of bioslurries into pulverized, coal-fired power stations is technically possible. The HHV of 1.4 t of bioslurry corresponds to about 1 t of hard coal; the subsidized German coal prices of *ca*. 300/t are almost comparable. Biomass ash and coal ash are different; this has to be considered, e.g., in case of fly ash use for cement production.

Bioslurry Gasification

Gasification Campaigns

The central step in the bioliq concept is the gasification of pumpable, char-, and ash-rich bioslurries from biomass pyrolysis in a PEF gasifier with a cooling screen. The feasibility of the concept was successfully verified in a 2- to 5-MW(th) PEF pilot gasifier built by Noell Company in 1996 for the further development of their hazardous waste conversion process [34]. The photo in Figure 22 shows the top and bottom sections of the pilot gasifier. The KIT has rented the pilot gasifier facilities from the successive owners (Babcock Borsig Power (BBP), Future Energy, Sustec Company; no campaign with the present owner of Siemens fuel gasification technology (FGT)) for four bioslurry gasification campaigns in each year from 2002 to 2005. For operation, the old, experienced crew from the previous DBI in Freiberg who developed the gasifier worked together with KIT personnel to transfer also some practical experience. The facilities are unfortunately not equipped for tests of the syngas cleaning and synthesis steps of the bioliq process.

Figure 22: A 3- to 5-MW(th) PEF (GSP-type) pilot gasifier with cooling screen in Freiberg, Germany (2005).

A total of 40 t of different bioslurries have been prepared by mixing 20 to 36 wt.% of either (1) milled charcoal with a typical LHV of 31 to 32 MJ/kg and an ash content of 2 wt.% or (2) pyrolysis char from cereal straw with a HHV of 25 to 26 MJ/kg and an ash content of about 15 to 20 wt.% into the following pyrolysis condensates: (a) a raw wood tar condensate from charcoal production from the Degussa process with a LHV of *ca.* 19 MJ/kg and a density of 1,160 kg/m^3; (b) an aqueous condensate, so-called 'raw wood vinegar,' with an LHV of 2 MJ/kg and a density of 1,030 kg/m^3; and (c) a condensate from intermediate straw pyrolysis [73] which had a LHV of 6 to 7 MJ/kg and is - as all other mainly aqueous condensates - not suited for direct gasification, only after addition of sufficient char powder to increase the HHV to at least 10 MJ/kg or more.

To generate a realistic straw slag layer at the inner surface of the gasification chamber, 3 wt.% of real straw ash - from the 3-MW(th) straw combustion facility in Schkölen, Germany - plus 0.3 wt.% KCl have been added to most slurries. Together with the ash content of the char, a total of about 4 wt.% ash has been present in the slurries. In the presence of about 50% SiO_2 and much potassium, typical ash melting points are below 1,200°C.

The following gasification conditions have been adjusted: a pressure of 26 bar and a temperature of 1,600°C to 1,200°C. A constant bioslurry feed rate of 0.35 to 0.6 t/h was maintained with a screw or plunger pump. The pressurized slurry stream was preheated to 40°C or 80°C by the immersion of the transfer pipe into hot water. Pressurized oxygen was generated by the evaporation of liquid O_2. In a special gasifier nozzle, the slow slurry stream was pneumatically atomized by a fast oxygen stream with a relative velocity of more than 100 m/s to supply sufficient energy for efficient atomization. To attain the high gasification temperature of 1,200°C or more, the oxygen mass flow must be almost half of the bioslurry mass flow (see Figure 23).

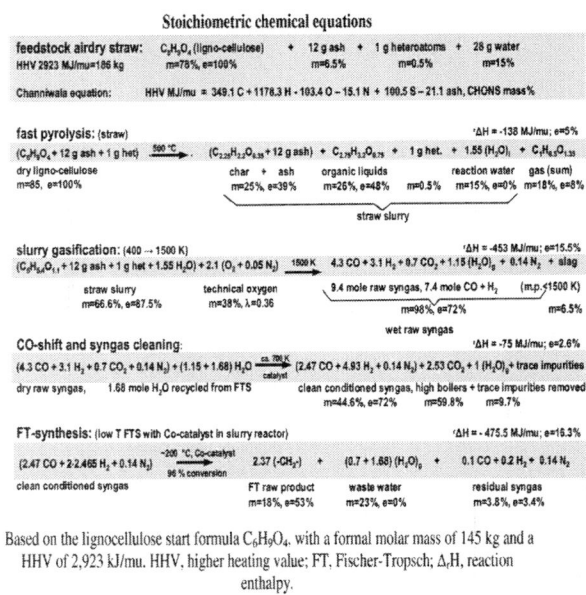

Figure 23: Set of coherent stoichiometric chemical equations for the successive steps in the bioliq process.

Because of the short gas residence time of only a few seconds in the gasification chamber and the low heat capacity of the membrane wall plus slag layer, new thermal stationary-state conditions after a change of the operation parameters have been attained in about an hour and have usually been kept for at least 2 h or more to characterize the stationary status by measurements. During each campaign, five to eight different stationary-state conditions have been studied. The slag residence time on the membrane screen is considerably higher, and after a change of the slag composition, it takes much more time in the range of 1 day until a new stationary slag layer has been built up. For the moment, slag layer equilibration problems have not been studied: In all our gasification experiments, we have kept the same straw slag composition, yet experiments with changing ash composition and longer and more expensive run times remain to be done.

The results of the many gasification runs in the pilot gasifier [74,75] are not presented in detail since the heat loss through the cooling screen is large because of the large surface-to-volume ratio in the relatively small, 5-MW gasifier and amounts up to about 20%. Corrections are complex and need extended explanation. The oxygen consumption in the pilot gasifier was found in the oxygen/fuel stoichiometry range of = 0.4 to 0.5. In a 100 times larger technical gasifier, the corrected stoichiometry ratio would be about 0.1 or 20% lower at values between 0.3 and 0.4.

Carbon conversion efficiency at higher temperatures has always been above 99%; even 99.8% have been calculated from the carbon measured in the slag and in the quench water. Above 1,200°C, the methane content in the syngas has been below 0.5% and usually even below 0.2% by volume. Among the potential tar constituents, only benzene could be found, in cooperation with the Paul Scherrer Institute, Switzerland, in amounts below 100 ppm. In general, the operation of the gasifier was smooth as expected and without surprise. Uncorrected experimental data from the pilot gasifier are compiled in Figure 24. The data are consistent with the expectation that the equilibrium of the CO-shift reaction is attained in the hot gasification chamber and maintained by the rapid total quench with an excess of cold water spray directly at the hot gasifier chamber exit. Depending on the detailed bioslurry composition, the dry raw syngas composition expected in a large technical PEF gasifier would be almost 60 vol.% CO and around 30 vol.% H_2, almost 10 vol.% CO_2 and few volume percent N_2. All essential gasification results with corrections for the screen loss are briefly summarized in Figure 25.

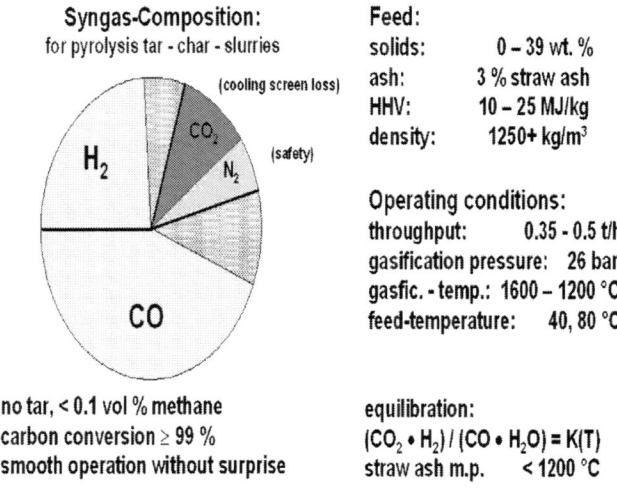

Figure 24: Gasification conditions and syngas composition for bioslurry gasification experiments in the Freiberg PEF pilot gasifier.

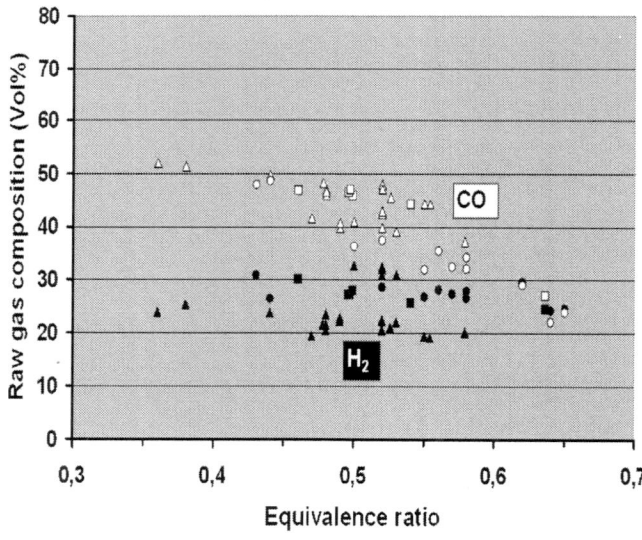

Figure 25: Corrected results from the PEF pilot gasifier without thermal screen loss corrections [75].

Modeling

Modeling a PEF gasifier in regard to the prediction of the product composition is relatively simple because at high gasification temperatures of ≥ 1,200°C and pressures of ≥ 25 bar, the thermodynamic equilibrium is quickly attained. Preconditions for a fast gasification are an efficient pneumatic bioslurry atomization and mixing with oxygen and a biochar particle size below 100 μm. At an oxygen injection velocity of > 100 m/s relative to the slurry, the kinetic energy is sufficient for efficient atomization and mixing. The oxygen demand for an overall exothermal gasification reaction at an adiabatic reaction temperature of 1,200°C can be reliably estimated from the stoichiometric gasification equation (see Figure 23) using tabulated thermodynamic data for the products. For bioslurries produced by FP of a dry lignocellulose, this results in a λ value of ca. 1/3, corresponding to a considerable oxygen mass of almost half the initial dry biomass.

Even at high pressures, a gasifier temperature above 1,200°C is sufficient to reduce thermodynamic CH_4 formation to tolerable low levels of < 1 vol.%. The high-temperature gasification conditions are also outside the soot formation limits. In the absence of soot, only the four CHO species: CO, CO_2, H_2O, and H_2, are coexisting in thermodynamic equilibrium of the homogeneous CO-shift reaction[76], a single key reaction which determines the composition of the hot raw syngas:

$$CO + H_2O \leftrightarrows CO_2 + H_2; \; K \; (T) = \frac{p(CO_2) \cdot p(H_2)}{p(CO) \cdot p(H_2O)} = \exp\left(4,578/T - 4.33\right); T \text{ in K}.$$

At high pressures and temperatures, equilibrium is quickly attained within seconds. The equilibrium of the homogeneous shift reaction is sufficient to characterize the thermodynamic situation in the gasifier chamber: With typical bioslurries, a dry raw syngas ($CO + H_2$) content above 80 vol.% (see Figure 23) is attained with H_2/CO ratios around 0.5 to 1.

Thermodynamic estimates are consistent with the experimental results of our gasification campaigns in the PEF pilot gasifier at Freiberg and also with general experience with PEF gasifiers. When the hot raw syngas is cooled down, the equilibrium methane content can increase, and the thermodynamic soot formation limit is crossed slightly below 1,000°C. Soot and methane formation have been prevented in our runs by a rapid total quench to ca. 180°C via injection of a water excess. Soot and methane formation became too slow, and the high-temperature equilibrium is frozen.

Further Operation and Development Aspects

Stationary-State Conditions

About 1 h after the start of the cold pilot gasifier, stable thermal conditions could be attained. After a change of the operating conditions, it took less than 1 h until new, stationary-state, thermal conditions are adjusted. This is true if the amount and the composition of the slag layer at the inner surface of the gasifier chamber do not significantly change. The formation of a completely new equilibrium slag coating is slower and takes several hours up to a day. The higher the ash content in the feed, the faster an equilibrium slag layer is formed. Effects of changing slag properties, e.g., by addition of fluxes, have not yet been investigated.

Operating Modes

Beside normal operation, start-up, shutdown, and standby operations deserve consideration. Within the huge and extended BTL complex, the gasifier section plays a special role. In the gasifier, an auxiliary burner for a clean natural gas or liquid fuel is needed. The auxiliary burner heats the cold gasifier to the desired temperature before the controlled gradual switch-over to a bioslurry feed takes place.

Auxiliary fuel is also required for shutdown and short-time standby operations. During a planned shutdown, the bioslurry feed in the piping system is replaced by clean liquid fuel. Thickness and composition of the slag layer in the gasifier are also adjusted. A sudden emergency shutdown is different and outlined in the 'Emergency' section.

In PEF gasifiers with a membrane wall (cooling screen), the heat capacity involved in temperature changes is low. The thick, heavy, and pressure-resistant outer mild steel cylinder behind the cooling screen has an almost constant temperature slightly below the 250°C to 300°C of the pressurized cooling screen water, and the mechanical stability remains high. The temperature drop and rise affects only the slag layer and the SiC liner at the inner surface of the membrane wall. Assuming a 2-cm-thick SiC liner plus a 2-cm-thick slag layer results in ca. 40 L or about 0.1 t of material/m^2. For a large PEF gasifier with a bioslurry plus an oxygen throughput of 0.15 t/s and a typical raw syngas residence time of 3 to 4 s, the stationary raw syngas mass in the gasification chamber amounts to less than 1 t. A PEF gasifier with an inner membrane wall surface of 100 m² in a ≥ 1-GW(th) gasifier has a slag mass inventory on the order of about 10 t. Heat-up can be attained in a start-up period of less than 1 h.

Emergency

A safety analysis for the whole BTL plant is complex, and here, only the gasifier is considered. The most credible major accident is the injection of cold oxygen or air into a reservoir of highly pressurized syngas. In a pressurized process, this is unlikely, but it can happen when the bioslurry feed into the gasifier suddenly stops or a drastic drop in the slurry heating value occurs and the oxygen flow continues. Countermeasure is an immediate stop of the oxygen inflow within a second. This must be securely guaranteed by several fast control measures with diverse methods.

Some design and operating characteristics of a large PEF gasifier contribute to safety. A multi-burner head in a large gasifier with several independent feed lines is unlikely to fail simultaneously. The extremely hot syngas ensures immediate oxygen combustion without accumulation of a combustible gas mix and subsequent explosion. The energy in the syngas inventory in the gasification chamber is limited and combustion would mainly cause a heat-up and meltdown of the large slag inventory. Gasifier operation with unstable biooils can be dangerous, e.g., if a plug of unobserved aqueous condensate with a low-heating value suddenly enters the gasifier. In dense slurries, the effects of phase separation are suppressed by the large amount of char present. An oxygen breakthrough into the cold syngas reservoirs below the ignition temperature downstream from quenching presents the major hazard. A sudden emergency shutdown is easily digested by the gasifier; parts of the slag liner will crack off from the SiC layer without causing damage, and the gasifier remains ready for immediate restart.

Heat Recovery from the Hot Syngas

The sensible heat in the 1,200°C hot raw syngas amounts to 15% to 20% of the initial bioenergy (see Figure 23), depending on the feed composition and the gasification temperature. Heat must anyway be removed from the raw syngas prior to cleaning. During syngas cooling, thermodynamic methane formation is favored, and the soot formation limits are crossed slightly below 1,000°C. Soot and methane formations have been suppressed to acceptably low levels by fast quench cooling to low temperatures with frozen equilibrium.

At temperatures above 600°C, volatile salts like KCl and entrained tiny and sticky slag (eutectics) droplets can cause technical problems. Large radiant heat exchangers can help but are expensive[77,78]. In our experiments, we have applied a simple total quench with excess water injection and recirculation. This option is technically simple and secure but causes the highest energy loss. Another option is total or partial quenching by recycling of cooled syngas; 500°C may be sufficiently low. Heat recovery from the hot raw syngas is not yet satisfactorily solved and needs further development.

Preheating of the Gasifier Feed

The reaction enthalpy of gasification must be sufficient to heat the products - mainly CO, CO_2, H_2, H_2O plus some slag - from ambient to *ca.* 1,200°C. If the feed is preheated, less oxygen is required, and the syngas yield increases correspondingly. In the pilot gasification experiments, the bioslurries were preheated at 26 bar to a maximum of 80°C using hot water. Pressurized bioslurries and oxygen can probably be preheated under pressure up to 150°C in a heat exchanger immediately prior to atomization in the gasifier using the abundant process waste heat. Even moderate preheating to 120°C can improve the syngas efficiency by more than 1%. The maximum temperatures which still guarantee sufficient short-time stability of the bioslurries have not been investigated in detail so far. The slurry decomposition limits during feed preheating are important and deserve further investigation.

Gasification Rates and Scale-Up

Precondition for fast and complete bioslurry gasification is an efficient mixing with oxygen. Gasification reactions proceed in the premixed turbulent downward gasifier flame in about a second and are completed by a few seconds of mixing the whole hot gasification chamber with the flame. The rate-controlling steps in a well-mixed flame are the heterogeneous gasification reactions of the porous char particles. This was confirmed by fluid dynamic simulations [79,80]. In the pilot gasification campaigns, it has been observed that char particles ≥ 0.2 mm escape unconverted into the slag.

At the high temperatures and pressures, the gasification rate is fast and proceeds in the film diffusion regime and is thus almost independent from temperature. The lower the temperature, the lower the oxygen consumption and the higher the syngas efficiency, at least until increasing methane formation or low gasification rates sets a lower limit. The lowest possible gasification temperature is therefore determined in most cases by a sufficiently low slag viscosity. Flux addition can be helpful to reduce the slag melting point to a reasonably low value. The large amounts of straw slag have usually sufficiently low melting points because of the high potassium contents. The smaller amounts of higher melting wood slag require either co-gasification with straw or flux addition.

In the film diffusion regime, the gasification rate increases with about the square root of pressure. A crude extrapolation from the 130-MW(th) GSP gasifier chamber at Schwarze Pumpe with a 2-m diameter and a 3.5-m height operating at 26 bar results in 1.8 GW(th) for a 4-m diameter and a 7-m height at an 80-bar operating pressure. A stepwise scale-up of the GSP gasifier to 0.5 GW(th) has already been realized by Siemens FGT in coal gasifiers in China, and further scale-up is intended. The heat loss through the membrane wall with values of 0.1 to 0.2 MW(th)/m² drops to below 1% for large-gigawatt gasifiers and becomes negligible.

Coal, Natural Gas, and Organic Waste as Feed

PEF gasifiers have already been operated with natural gas, pulverized coal, or coal dust/water slurries. Large GTL and CTL plants are already in operation and can be combined with BTL plants to flexible multi-feed x-to-liquid (XTL) facilities in the future. During a transition period with gradually decreasing crude oil supply, development and market introduction of XTL synfuel plants is a reasonable and precautional strategy.

Syngas Cleaning, Conditioning, and Use

Raw Syngas Cleaning and H_2/CO Ratio Adjustment

Biosyngas cleaning procedures for bioliq have not yet been investigated so far on a larger scale but are included in the future bioliq R&D and pilot plant development program [42,81]. Not only the conventional syngas cleaning options will be considered, but also special hot gas cleaning technologies as under construction in the pilot plant. Another special task is the investigation of possibilities to eliminate the technically complex and expensive CO-shift reaction by integration of the shift into the final synthesis. As an example, the DME synthesis is described in the following 'DME synthesis' section.

With an initial $C_6H_9O_4$ composition for dry lignocellulose and a temperature of 1,200°C in a PEF gasifier, a raw syngas with a low H_2/CO volume ratio of 0.5 to 0.7 is generated (see Figure 23). A H_2/CO volume ratio of about 2 is optimum for methanol synthesis with Cu catalysts or FT synthesis with Co catalysts; about half of the CO must therefore be converted to H_2. For DME synthesis, a H_2/CO ratio of 1 is sufficient. Without the technically complex and expensive CO-shift step, raw syngas cleaning would be limited to the much cheaper removal of undesired trace constituents, acting as catalyst poisons.

DME Synthesis

Among the many syngas applications, we have focused on DME synthesis for several reasons: DME is a clean fuel (b.p. -25°C) and a neat diesel fuel in particular; handling is similar to LPG. Another direct application is as a low-toxicity propellant. Its use as an important intermediate and platform chemical for the further conversion to various chemicals and fuels is expected to gain importance in the future. Methanol is a similar 'platform chemical,' and most methanol conversion reactions proceed via DME as an intermediate. Methanol has the advantage that it can be stored and transported in large amounts as a liquid at atmospheric pressure. However, for the conversion into synfuels, e.g., in the MtS process [45], large production capacities are needed without the necessity of large storage and transport requirements for an intermediate. The production route can proceed directly to the final product without process interruption and large intermediate buffers. In such cases, it is reasonable to convert syngas directly to DME and not via a methanol intermediate because of the following reasons:

- A much higher thermodynamic synthesis yield for DME can be obtained in a single pass through the catalyst bed (see Table 6) [82]. Methanol synthesis requires much expensive recycle of unconverted syngas.

Table 6: Key reactions for DME synthesis from syngas

Reaction	Reaction equation	$\Delta_r H$ (kJ / mol)	ΔrG (kJ / mol)
1. CH_3OH synthesis	$2\ CO\ +\ 4\ H_2 \rightleftarrows 2\ CH_3OH_{gas}$	-182	-58
2. CH_3OH dehydratization	$2\ (CH_3OH)_{gas} \rightleftarrows DME\ +\ (H_2O)_{gas}$	-24	-4
3. CO-shift reaction	$CO\ +\ (H_2O)_{gas} \rightleftarrows H_2 +\ CO_2$	-41	-29
4. DME synthesis	$3\ CO\ +\ 3\ H_2 \rightleftarrows DME\ +\ CO_2$	-247	-91

- DME is usually produced in a two-step process: (1) methanol from syngas and (2) methanol dehydration. The conditions for both reactions can be selected independently from each other. A one-step process at mutually harmonized synthesis conditions offers economic advantages especially for large-scale production.

- A highly selective $Cu-ZnO-Al_2O_3$ methanol catalyst acts also as a low-temperature CO-shift catalyst at the usual methanol or DME synthesis conditions. If a dehydration catalyst, e.g., $\gamma-Al_2O_3$ or a zeolite (e.g., HZSM-5), is admixed for methanol dehydration to DME, the reaction water will be consumed on-site in the CO-shift reaction. In a dry syngas, this allows a reduction of the H_2/CO ratio to at least 1; in a moist syngas, it may be even lower. It would be a big economic advantage if the technical effort for a CO shift plus downstream CO_2 removal in the syngas cleaning and conditioning train is not needed.

The three simultaneous key reactions 1, 2, and 3 in the catalyst bed are given in Table 6, together with their $\Delta_r H$ and $\Delta_r G$ values. These thermodynamic data show the highly exothermal $\Delta r H$ and the equilibrium Gibbs free energy $\Delta r G = -RT \ln K$, but no details of the rather complex heterogeneous reaction kinetics and mechanisms. Catalyst design must therefore rely on experimental data. For the sum reaction 4, it can be calculated that the equilibrium for a H_2/CO ratio of 1 at 250°C and a 50-bar pressure is at 95% conversion to DME [82]. This seems to be high enough for a single pass conversion without recycle, at least in a two-stage synthesis reactor with intermediate DME removal.

The considerations above have been successfully realized in a lab-scale facility shown in Figure 26. The bifunctional catalyst was a mix of 1 or 1.5 g of a commercial $Cu/ZnO/Al_2O_3$ methanol catalyst and 1 or 0.5 g of $-Al_2O_3$ as a dehydration catalyst, with a grain size of ca. 0.5 mm and a volume of ca. 2 ml. Operating pressure and temperature ranges were 25 to 50 bar and 200°C to 275°C, respectively; catalyst deactivation starts at 280°C by the sintering of the methanol catalyst. The gas flow was 3.3 to 6.5 NL/h; the syngas H_2/CO ratios were 2, 1, and 2:3. The syngas has been diluted with 70 vol.% Ar to guarantee isothermal conditions in the small, thermostatted reactor. Catalyst activation procedures were performed according to manufacturer's instructions.

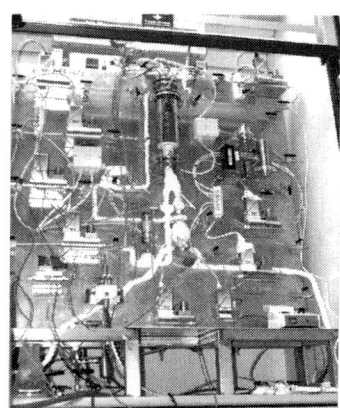

Figure 26. Laboratory synthesis device (0.2 m³/h, 300°C, 60 bar).

Suitable DME synthesis conditions at a H_2/CO ratio of 1 have been found at 250°C and 50 bar. Both methanol and water vapor concentrations in the product gas have always been low and near the detection limit [46,47]. This confirms the above mentioned considerations that dehydration and CO shift are fast reactions. The investigations have been extended using different catalysts, temperatures, space/time velocities, and H_2/CO ratios [48].

Economic Aspects

The major hurdle for biosynfuel introduction into the market is the higher price compared to untaxed, oil-derived motor fuels, though synfuels have higher purity and quality and can be tailored for innovative internal combustion engines. Recent cost estimates for the bioliq BTL process [83-85] are summarized in the following.

Mass and Energy Balance

Indispensible bases for an economic analysis and modeling are reliable mass and energy balances. For the sequence of chemical reactions in the bioliq process, our experimental results and complementary literature data have been condensed into a coherent set of empirical chemical stoichiometric equations. All equations in Figure 23 are based on a water-, ash-, and heteroatom-free lignocellulose start 'molecule' with the formula $C_6H_9O_4$, formal molecular mass = 145 kg/mu, combustion heat $\Delta_c H$ = -2,923 MJ/mu. The usual thermodynamic tools can be applied to these chemical equations. To represent the reality, moisture, ash, (nonvolatile inorganics) and heteroatoms (e.g., N, S; volatile, non-CHO elements) have been added to the organic lignocellulose fraction. Dry lignocellulose from wood and straw differs in good approximation only in the ash content of about 1 or 6 wt.% on the average. Mass (m) and energy (e) percentages are given for each reaction partner in the chemical conversion train and refer to the initial biofeedstock: air-dry cereal straw with m = 100% and e = 100%. This allows a convenient and quick overview on the mass and energy flows in the process. Successive mass streams are 7 t of air-dry straw (15 wt.% H_2O, LHV 4 kWh/kg) → 6 t of dry straw → ca. 4.7 t of biosyncrude (slurry or paste LH ca. 5.4 kWh/kg) → 1.25 t of FT raw product (LHV ca. 12 kWh/kg) → 1 t of biosynfuel (LHV 12 kWh/kg).

The standard reaction enthalpies are the sum over the combustion heats (HHV) of all individual reactants i: $\sum \Delta_c H_i$. Unknown combustion heats for organics have been estimated with fair ± 1.5% accuracy from their CHO (N, S, ash) composition with the Channiwala equation. Since this equation is linear in the elements, a zero $\Delta_r H$ will result if the Channiwala relation is applied to all reactants. Typical outliers are the small molecules like the combustion products H_2O and CO_2 or pure carbon as shown in Table 4. Thus, an estimate of $\Delta_r H$ is only possible if the yields of these small molecules are known or have been measured. The reaction enthalpy is then the sum over these yields times the $\Delta_c H$ difference from the Channiwala equation and measured $\Delta_c H$ data.

For FP, the experimental sum of combustion heats of the products is usually found to be lower than that of the educts, indicating an exothermal reaction. However, since some volatile oxygenates with low molar mass escaped unobserved and unanalyzed with the pyrolysis gases - noticed not only in our experiments - this is a reasonable explanation for the deficit of several percent usually observed in the experimental mass and energy balances from biomass FP. The measured average heat requirements of ca. 1.3 MJ/kg for FP of dry lignocellulose [35,66] are higher than the value of about 0.9 MJ/kg expected for heating-up the FP products from ambient to 500°C; this indicates a slightly endothermal FP with positive $\Delta_r H$. The lowest observed heat requirement of around 0.9 MJ/kg is comparable with the energy required to heat the pyrolysis products from ambient to 500°C. This is a hint that the FP reaction of lignocellulose is occasionally thermoneutral.

In practice, ca. 10% of the initial bioenergy is required for FP without heat recovery during quench condensation and including some heat losses in the hot loop section. Not included is the energy for biomass diminution and electricity consumption, e.g., for cooling water and process pumps; this requires several percent more. Energy invested into FP is at least partly compensated by energy savings in downstream processing, e.g., by the significant mass reduction from 100 wt.% of biomass feed to about 80 wt.% bioslurry at a 20 wt.% pyrolysis gas yield.

The energy flow diagram in Figure 27 shows that the chemical combustion energy $\Delta_c H$ in the successive products decreases stepwise due to the liberation of energy as reaction heat: the higher the reaction temperature, the more valuable is this thermal reaction energy if conversion to mechanical or electric power is considered. Not more than about 40% to 50% of the initial bioenergy can be converted into a final biosynfuel, e.g., via the FT process. The conversion efficiency of biomass-to-synfuel depends much on by-product management. Recycle of unconverted syngas and by-products improves the product yields but is usually more expensive than the simpler direct combustion to generate energy, especially for a self-sustained process. A self-sustained process without import or export of energy is assumed to be near the economic design optimum.

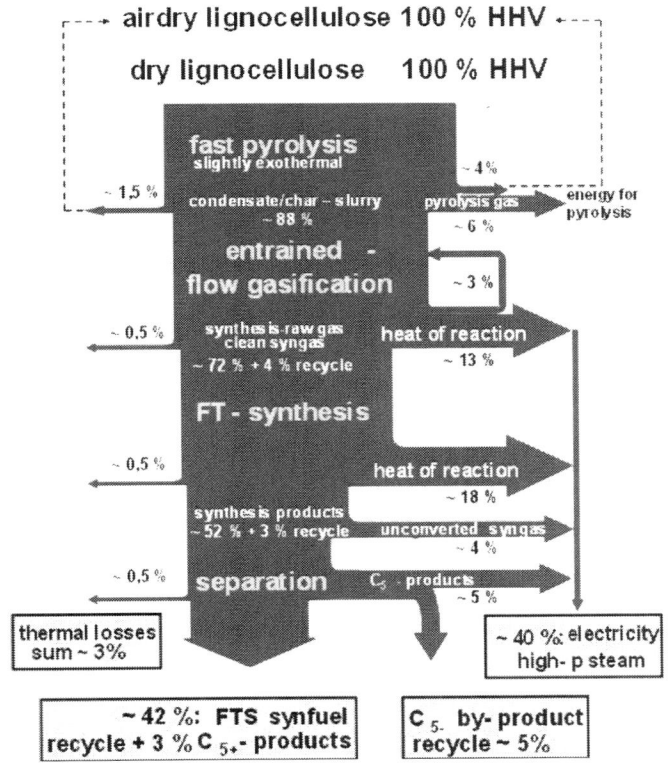

Figure 27: Energy flow in the successive steps of the bioliq process (based on Table 6).

General Economy Aspects.

Manufacturing costs of biosynfuels are affected by some general plant characteristics.

- *Plant capacity.* The TCI for FP or BTL-type chemical facilities does not increase linearly with capacity but with a cost degression exponent of *ca.* 0.7:

$$\frac{\text{cost (new)}}{\text{cost (reference)}} = \left[\frac{\text{capacity (new)}}{\text{capacity (reference)}} \right]^{0.7};$$

TCI for a ten times higher production rate increases then only by a factor of 5; in other words, a capacity increase by an order of magnitude reduces the specific TCI to about half. As a consequence, BTL plants or thermochemical biorefineries should be as large as reasonably possible. A modern oil refinery produces about 10 Mt/a of oil-derived motor fuels. The production capacity of a biosynfuel plant should be at least 1 Mt/a or 10% of an oil refinery for competition reasons. A reasonable size for the much smaller regional FP plants is a maximum delivery distance of *ca.* 30 km for the local farmers with their tractors. It is an accepted experience from the seasonal sugar beet transport that a faster transport with trucks reduces the local traffic holdup without much transport cost difference.

A regional 100-MW(th) FP plant with *ca.* 200 kt of annual biomass throughput (25 t/h of air-dry lignocellulose with 15 wt.% H_2O and LHV of 4 kW/kg in 8,000 h) requires about 13,000 truck transports/year with a 100-m³ volume and 15-t straw bale mass on the average of 30 km back and forth. Delivery during 3,000 h/a daylight without weekends and holidays requires at least a dozen of special trucks in permanent operation.

- *Biomass transport costs.* In central EU, about 1 Mt/a of residues from forestry and agriculture (half the cereal straw harvest) is available within a radius of about 70 km. Because of the tortuous roads, this transportation distance is also a reasonable estimate for the mean transport distance. Above about 70-km distance, direct truck delivery of air-dry straw to a central plant becomes more expensive than regional pyrolysis plants followed by biosyncrude or bioslurry transport in silo wagons with the electrified rail to a large central BTL plant.

A BTL plant with 1-Mt/a biosynfuel capacity is a reasonable minimum size and needs *ca.* 7 Mt/a of air-dry wood or straw. The local traffic density for truck delivery to such a huge plant would be an unacceptable nuisance for the local population and requires additional money to extend and maintain the local road infrastructure. A 100-m³ truck load with 15 t of straw bales would result in a truck movement in and out every half minute without interruption for days, nights, and weekends.

- *'Brownfield'* or *'grass root'* plant site. A plant site within an already existing industrial complex (brownfield), e.g., an oil refinery or a chemical complex, enables remarkable cost savings since many auxiliary facilities are already available. In a grass root or 'greenfield' site, additional money is required for the erection of these facilities. Rail access is of special importance for a biorefinery because of the large delivery volumes.

Cost reduction by learning. Capex and opex decrease with increasing practical experience. An exponential cost decrease is usually assumed for successive plants, down to a minimum of maybe only half of the TCI for the very first commercial plant. This will be a fast process for the large number of FP plants required; for the huge BTL complexes, a much slower stepwise decrease is expected. Experience from the operation of the technically similar CTL and GTL plants will contribute a lot to the learning process, especially in the tail-end section. Also, a common operation with different feed materials can be considered in huge XTL plants. The stepwise cost decrease with growing experience can be complemented by cost savings by simultaneous orders of equipment or by replication if several plants of the same design are erected in the so-called convoy mode. This also reduces engineering costs.

Biosynfuel Manufacturing Cost Estimation

The cost estimation follows standard methods described in textbooks [86-89]. The following five cost contributions have been considered:

- Raw materials (degression exponent = 1)
- Utilities (zero cost assumed for a self-sustained process)

- Labor and related costs (degression exponent = 0.25)
- Capital and related costs (degression exponent = 0.7)
- Generalia (*ca.* 6% from 1 + 2 + 3 + 4 sum assumed).

Capital and capital-related costs in this script do not refer to the very first commercial plant but to a brownfield plant with mature technology in more than a decade from now. Annual capital and related costs are assumed to be 25% of the TCI for a new plant. The GTL plant Oryx-1 in Qatar with a capacity of 35,000 bbl/day and a TCI of $1.1 billion, equivalent to about €1 billion at the report date, was the basis of the estimate. The scaled TCI of $0.75 billion for the slightly smaller bioliq BTL plant has been estimated without the costs for central biomass preparation by FP and without an air separation unit (ASU); FP and ASU together make up *ca.* 25% of TCI!

The higher oxygen demand in a BTL plant compared to the gasification of coal or natural gas was considered by oxygen delivery over the fence as a raw material. Figure 27 shows that a large bioenergy fraction of almost half becomes available for heat, high-pressure steam, or electricity generation. A self-sustained process including the large electricity consumption in the ASU and negligible additional utility costs have been assumed. The TCI of a single central bioliq BTL plant without FP preparation for 1 Mt/a of biosynfuel is comparable to the TCI for about 38 FP plants. Together, they add to a total investment of 0.75 + 0.76 G€ or about €1.5 billion (see Table 7). An economic assessment of different FP plants is discussed in the study of Peacocke et al. [90].

Table 7: Manufacturing cost contributions of FT raw syndiesel

Bioslurry manufacturing	Capacity	Synfuel production	
		Cost (/t)	%
a. Regional biosyncrude production in FP plants with a 200-kt/a TCI of 20 to 25 M€/plant for mature technology			
1. Raw materials: air-dry straw on the field	7 t; €45/t	315	28.9
2. 30-km tractor transport of straw bales	7 t; €18/t	126	11.6

3. Biosyncrude production costs by FP	4.7 t; €37.3/t	175	16.1
4. Labor: 25 persons/plant à 60 k€/a		53	4.9
Sum		669	61.5
b. Central biosyncrude gasification and FT synthesis for 1 Mt/a of biosyndiesel (TCI without ASU *ca.* €0.75 billion, scaled from Oryx-1)			
1. 250-km biosyncrude transport by rail	4.7 t; €21/t	69	6.3
2. 1,400 m³ (STP) oxygen ± €0.06/m³ (STP) (O_2 price without electricity)		84	7.7
3. Biosyncrude gasification and FT synthesis		188	17.2
4. Labor: 300 persons à 60 k€/a		18	2
Sum		359	33.2

FP, fast pyrolysis; TCI, total capital investment; FT, Fischer-Tropsch; ASU, air separation unit; STP, standard temperature and pressure.

Dahmen et al. Energy, Sustainability and Society 2012 2:3 doi:10.1186/2192-0567-2-3

Average personnel costs of 60 k€/person/year are a reasonable assumption for central EU. A price of about €60 to €70/t for residual air-dry (15 wt.% H_2O) straw or wood delivered to the FP plant yard (free on board (f.o.b.)) is also reasonable [6]. One should not forget that much of the biomass f.o.b. costs are labor costs or new jobs in agriculture. The individual cost contributions are summarized in Table 7 and visualized in Figure 28. The numbers in Table 7 can easily be adjusted to other conditions or replaced if better estimates or assumptions are available.

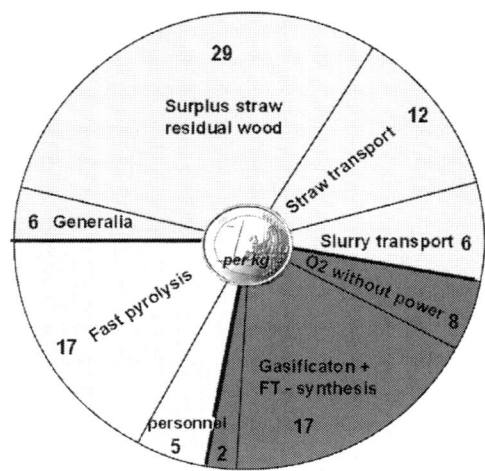

Figure 28: Breakdown of the biosynfuel manufacturing costs for the Karlsruhe bioliq process.

The total manufacturing costs for 1 t of biosyndiesel raw product with 6% generalia, but without biosynfuel distribution costs and profit, amount to *ca.* €1,090/t or (at a syndiesel density of 750 kg/m³) *ca.* €820/ m³ with a large uncertainty on the order of ± 30%. This is about 1.5 times the known cost for untaxed, oil-derived diesel at an oil refinery gate and at the present crude oil price of *ca.* €80/bbl ($100/bbl) mid of 2011.

The manufacturing cost breakdown for biosyndiesel in Figure 28 shows that the bio-feedstock plus transport causes about half of the costs. The other half consists of technical costs, which are about proportional to the specific TCI and decrease with plant capacity. In countries with low biomass costs, biosynfuel competition with oil-derived fuels will be attained much earlier. Labor costs do not play a decisive role in the large plants. Other methods for cost estimates give similar results [6] and remain within the given error range of ± 30%.

CONCLUSIONS AND OUTLOOK

Biomass is the only renewable carbon resource and will gradually become the major raw material for organic chemistry. Large physicochemical, biochemical, and thermochemical biorefineries are the organic chemical industries of the future and produce organic chemicals, biosynfuels, plus various forms of energy as inevitable couple and side products. The thermochemical pathway proceeds via the conversion of the abundant lignocellulosic biomass to syngas and allows a rather flexible use not only of various biofeedstock, but also of coal and other fossils during a transition period.

The bioliq process under development at the KIT can be considered as the backbone of a large future thermochemical biorefinery. Lignocellulosic biomass is first liquefied by FP to a bioslurry. This is a convenient handling, storage, and transport form and has been successfully gasified in a slagging PEF gasifier to a tar-free, low-methane syngas. The front-end steps for syngas generation from biomass still need further development to achieve commercial maturity. The tail-end steps for syngas use are well known and practiced technically since almost a century. A vast literature exists for syngas generation and use. The feasibility of the essential new steps of biomass liquefaction, bioslurry preparation, and gasification in a PEF gasifier has been successfully verified in a number of lab- and pilot-scale experiments.

AUTHORS' CONTRIBUTIONS

FW as the chief engineer of the pyrolysis facilities at KIT carried out most of the experimental work on fast pyrolysis. EH along with ED worked out the bioliq concept; EH also performed the principles of process design and estimations. ND as the bioliq project manager added the relevant text and drafted the manuscript together with EH. All authors read and approved the final manuscript.

ACKNOWLEDGEMENTS

We appreciate the substantial financial support from the German Ministry of Food, Agriculture and Consumer Protection, the Agency of Renewable Resources, the state of Baden-Württemberg, and the EU Commission for funding the RENEW and BioBOOST projects.

REFERENCES

1. Berié Eva (2009) Der Fischer Weltalmanach 2010: Zahlen Daten Fakten. Fischer-Taschenbuch Verlag: Frankfurt.

2. Kojima T (1998) The carbon dioxide problem: integrated energy and environmental policies for the 21st century. Overseas Publisher Association, UK.

3. Dostrovsky I (1988) Energy and the missing resource: a view from the laboratory. Cambridge University Press, UK.

4. Doran M (2009) Contribution of energy crops. In: Bridgwater AV, Hofbauer H, van Loo S (eds) Fast pyrolysis of biomass: a handbook, CPL Press, UK.

5. Seyfried F (2011) Renewable fuels for advanced powertrains (RENEW): executive summary. [http://www.renew-fuel.com/home.php] Accessed 25 Aug 2011

6. Leible L, Kälber S, Kappler G, Lange S, Nieke E, Proplesch P, Wintzer D, Fürniβ B (2007) Kraftstoff, Strom und Wärme aus Stroh und Waldrestholz: eine systemanalytische Untersuchung. FZK, Karlsruhe.

7. Klass DL (1998) Biomass for renewable energy, fuels, and chemicals. Academic Press, San Diego.

8. Dahmen N, Henrich E, Kruse A, Raffelt K (2010) Biomass liquefaction and gasification. In: Vertès AA, Qureshi N, Blaschek HP, Yukawa H (eds) Biomass to biofuels: strategies for global industries, Wiley Blackwell Science, UK.

9. Kamm B, Gruber PR, Kamm M (eds) (2006) Biorefineries - industrial processes and products. Wiley-VCH, Weinheim.

10. Skrzypek J, Sloczynski J, Ledakowicz S (1994) Methanol synthesis: science and engineering. Polish Scientific Publishers, Warsaw.

11. Olah GA, Goeppert A, Prakash GKS (2006) Beyond oil and gas: the methanol economy. Wiley-VCH, Weinheim.

12. Schaub G, Unruh D, Rhode M (2004) Synfuels from biomass via Fischer-Tropsch-synthesis - basic process principles and perspectives. Erdöl Erdgas Kohle 120:327-331 PubMed Abstract |

13. Wender I (1996) Reaction of synthesis gas. Fuel Processing Technology 48:189-297

4. Ertl G, Knözinger H, Weitkamp J (1997) Handbook of heterogeneous catalysis. Wiley-VCH, Weinheim.

15. Winnacker-Küchler (1981) Organische Verfahren auf der Basis von Synthesegas. Chemische Technologie 5:502-588

16. Higman C, van der Burgt M (2003) Gasification. Elsevier Science, Burlington.

17. Hofbauer H (2009) Gasification-technology overview. In: Bridgwater AV, Hofbauer H, van Loo S (eds) Fast pyrolysis of biomass: a handbook, CPL Press, UK.

18. Henrich E, Dinjus E, Meier D (2002) Flugstromvergasung von flüssigen Pyrolyseprodukten bei hohem Druck: Ein neues Konzept zur Biomassevergasung. Tagungsbericht 2002-2: Energetische Nutzung von Biomasse, Velen V, Velen.

19. Henrich E, Dinjus E, Weirich F (2002) A new concept for biomass gasification at high pressure. In: Palz W, Spitzer J, Maniatis K, Kwant K, Helm P, Grassi A (eds) Proceedings of the 12th European conference and technology exhibition on biomass for energy, industry and climate protection, Amsterdam ETA, Florence and WIP, Munich, vol 1.628-632

20. Henrich E, Dinjus E (2003) Tar-free, high pressure syngas from biomass. In: Bridgwater AV (ed) Pyrolysis and gasification of biomass and waste: proceedings of an expert meeting, Strasbourg, September-October 2002, CPL Press, UK. pp 511-526

21. Henrich E, Dinjus E (2004) Das FZK-Konzept zur Kraftstoffherstellung aus Biomasse. In: FNR (ed) Biomasse-Vergasung - Der Königsweg für eine effiziente Strom- und Kraftstoff- Bereitstellung? Schriftenreihe Nachwachsende Rohstoffe, Band 24. Landwirtschaftsverlag, Münster, pp. 298-337

22. Henrich E, Weirich F (2004) Pressurized entrained flow gasifiers for biomass. Environmental Engineering Science 21:53-64

23. Henrich E (2004) A two-stage process for synfuel from biomass. In: van Swaaij WPM, Fjällström T, Helm P, Grassi A (eds) Proceedings of the 2nd world conference and technology exhibition on biomass for energy, industry and climate protection, Rome ETA, Florence and WIP, Rome, vol 1.729-733

24. Henrich E, Weirich F (2004) A twin screw mixer reactor for fast pyrolysis of biomass. In: van Swaaij WPM, Fjällström T, Helm

P, Grassi A (eds) Proceedings of the 2nd world conference and technology exhibition on biomass for energy, industry and climate protection, Rome ETA, Florence and WIP, Rome, vol 1.982-985

25. Henrich E, Dinjus E, Meier D (2004) Synthesegas aus verflüssigter Biomasse. Tagungsbericht 2004-1. Energetische Nutzung von Biomasse Velen VI, Velen.

26. Henrich E, Leible L, Malcher L, Wiemer HJ (2005) Gaserzeugung aus Biomasse. Project report (Ministry of Agriculture of the state of Baden-Württemberg, Germany.ref. no. 46(54)-8214.07).

27. Henrich E, Raffelt K, Stahl R, Weirich F (2006) Clean syngas from biooil/char-slurries. In: Bridgwater AV, Boocock DGB (eds) Science in thermal and chemical biomass conversion, vol 2. CPL Press, UK, pp 1565-1579

28. Dahmen N, Henrich E (2007) Synthesis gas from biomass. Oil Gas European Magazine 22:31-34

29. Schingnitz M (2008) Flugstromvergasung, Gaskombinat Schwarze-Pumpe-Verfahren (GSP). In: Schmalfeld J (ed) Die Veredelung und Umwandlung von Kohle, DGMK, Hamburg.

30. Seifert W, Buttker B (2000) Stoffliche Abfallverwertung über die Route Vergasung, Synthesegasherstellung, Methanolgewinnung in der SVZ Schwarze Pumpe. Tagungsbericht 2000-1. Energetische Nutzung von Biomasse Velen I, Velen.

31. Schingnitz M (1994) Darstellung des Noell-Konversionsverfahrens. In: Carl J, Fritz P (eds) Noell Konversionverfahren zur Verwertung und Entsorgung von Abfällen, EF-Verlag, Berlin.

32. Schingnitz M (2002) Die Vergasungstechnik-Eine Chance zur Gestaltung umweltkompatibler Prozesse in der chemischen und Zellstoffindustrie. Chemie-Ingenieur Technik 74:776

33. Schingnitz M, Volkmann D (2004) Sind Erfahrungen aus der Flugstromvergasung von konventionellen Brennstoffen und Abfallstoffen auf Biomasse übertragbar. Tagungsbericht 2004-1. Energetische Nutzung von Biomasse Velen VI, Velen.

34. Carl J, Fritz P (eds) (1997) Noell Konversionsverfahren zur Verwertung und Entsorgung von Abfällen. EF-Verlag, Berlin.

35. Kornmayer C (2009) Verfahrenstechnische Untersuchungen zur Schnellpyrolyse von Lignocellulose im Doppelschnecken-Mischreaktor. PhD thesis, Karlsruhe University (TH).

36. Bridgwater AV, Peacocke GVC (2000) Fast pyrolysis processes for biomass. Renewable and Sustainable Energy Reviews 4:1-73

37. Raffelt K, Henrich E, Steinhardt J (2004) Herstellung und Eigenschaften von Slurries aus Biomasse für die Biomassevergasung. Tagungsbericht 2004-1. Energetische Nutzung von Biomasse Velen VI, Velen.

38. Raffelt K, Henrich E, Kögel A, Stahl R, Steinhardt J, Weirich F (2004) Stable slurries from biomass pyrolysis products for entrained flow gasification. In: van Swaaij WPM, Fjällström T, Helm P, Grassi A (eds) Proceedings of the 2nd world conference and technology exhibition on biomass for energy, industry and climate protection, Rome ETA, Florence and WIP, Munich, vol 1.914-917

39. Raffelt K, Henrich E, Steinhardt J, Dinjus E (2006) Preparation and characterization of biomass slurries: a new feed for entrained flow gasification. In: Bridgwater AV, Boocock DGB (eds) Science in thermal and chemical biomass conversion, vol 2. CPL Press, Newbury, pp 1547-1558

40. Reschke AE (1998) The development of colloidal mixer based CRF systems. Bloss M Carlton (ed) MINEFILL'98. 6th international symposium on mining with backfill, Brisbane.

41. Doran M (2009) Feedstocks for thermal conversion. In: Bridgwater AV, Hofbauer H, van Loo S (eds) Fast pyrolysis of biomass, CPL Press, UK.

42. Leibold H, Seifert H (2010) Energetische Aspekte bei der Heißgasreinigung von biomassebasierten Synthesegasen. In: Thome-Kozmiensky KJ (ed) Erneuerbare Energien, vol 4. TK Verlag, Berlin, pp 95-107

43. Dinjus E, Arnold U, Dahmen N, Höfer R, Wach W (2009) Green fuels-sustainable solutions for transportation. In: Höfer R (ed) RCS Green Chemistry No. 4: sustainable solutions for modern economies, RCS Publishing, London.

44. Thomas JM, Thomas WJ (1997) Heterogeneous catalysis: examples and case histories. In: Boatman D (ed) Principles and practice of heterogeneous catalysis, VCH, Weinheim.

45. Liebner W, Wagner M (2004) Mt synfuels, the efficient and economical alternative to Fischer-Tropsch fuels. Erdöl Erdgas Kohle 120:323-326 PubMed Abstract |

46. Mas C, Dinjus E, Ederer H, Henrich E, Renk C (2006) Dehydratisation of methanol to dimethylether. FZK, Karlsruhe.

47. Renk C (2009) Die einstufige Dimethylether-Synthese aus Synthesegas. PhD thesis, Karlsruhe University (TH).

48. Stiefel M, Ahmad R, Arnold U, Döring M (2011) Direct synthesis of dimethyl ether from carbon-monoxide-rich synthesis gas: influence of dehydration catalysts and operating conditions. Fuel Processing Technology 92:1466-1474

49. Dahmen N, Dinjus E (2010) Das bioliq-Verfahren-Konzept, Technologie und Stand der Entwicklung. Motortechnische Zeitschrift (MTZ) 71:864-868

50. Dahmen N, Dinjus E, Kolb T, Arnold U, Leibold H, Stahl R (2012) State of the art of the bioliq®process for synthetic biofuels production. In: Abraham M (ed) Environmental progress & sustainable energy, Wiley, Hoboken (in press).

51. Scott DS, Piscorz P (1982) The flash pyrolysis of aspen poplar wood. Can J Chem Eng 60:666-674

52. Scott DS, Piscorz P, Radlein D (1985) Liquid products from the continuous flash pyrolysis of biomass. Ind Eng Chem Process Des 24:581-588

53. Bridgwater AV, Czernik S, Diebold J (eds) (2005) Fast pyrolysis of biomass: a handbook. CPL Press, UK.

54. Czernik S, Bridgwater AV (2004) Overview of application of biomass fast pyrolysis Oil. Energy & Fuels 18:590-598

55. Bridgwater AV (2009) Fast pyrolysis of biomass. In: Bridgwater AV, Hofbauer H, van Loo S (eds) Fast pyrolysis of biomass, CPL Press, UK.

56. Wagenaar BM, Venderbosch RH, Carrasco J, Strenzick R, van der Aa BJ (2001) Rotating cone biooil production and applications. In: Bridgwater AV (ed) Progress in thermochemical biomass conversion, Blackwell Science, Oxford.

57. Prins W, Wagenaar BM (1997) Review of rotating cone technology for flash pyrolysis of biomass. In: Kaltschmitt MK, Bridgwater AV (eds) Biomass gasification and pyrolysis, CPL Press, UK.

58. Schmalfeld J (2008) Pyrolyse, Lurgi-Ruhrgas-Verfahren (LR). In: Schmalfeld J (ed) Die Veredelung und Umwandlung von Kohle, DGMK, Hamburg.

59. Yang J, Blanchette D, de Caumia B, Rey C (2001) Modeling, scale-up and demonstration of a vaccum pyrolysis reactor. In: Bridgwater AV (ed) Progress in thermochemical biomass conversion, Blackwell Science, Oxford.

60. Meier D (2005) New ablative pyrolyser in operation in Germany. In: PyNe newsletter 17. Aston University, UK.

61. Weiming Y, Xueyuan B, Fang H, Zhihe L, Yongjun L, Shuangning X (2004) Biomass pyrolysis liquefaction technology. Proceedings of the international conference on biomass pyrolysis liquefaction technology, Beijing.

62. Peters W (1963) Schnellentgasung von Steinkohlen. In: Habilitation thesis. RWTH Aachen University.

63. Channiwala SA, Parikh PP (2002) A unified correlation for estimating HHV of solid, liquid and gaseous fuels. Fuel 81:1051-1063

64. Dahmen N (2008) Die Schnellpyrolyse im Rahmen des bioliq®-Verfahrens am Forschungszentrum Karlsruhe. In: FNR (ed) Schriftenreihe Nachwachsende Rohstoffe Band 28, Bio crude oil. Landwirtschaftliche Verlagsanstalt, Münster,

65. Riegel ER (1933) Industrial chemistry. The Chemical Catalog Company, Inc., New York.

66. Daugaard DE, Brown RC (2003) Enthalpy for pyrolysis of several types of biomass. Energy & Fuels 17:934-939 PubMed Abstract |

67. Antal MJ, Gronli M (2003) The art science and technology of charcoal production. Ind Eng Chem Res 42:1619-1640

68. Pawlowski P (1971) Die Ähnlichkeitstheorie in der physikalisch technischen Forschung. Springer Verlag, Berlin.

69. Raffelt K, Dahmen N, Dinjus E, Henrich E, Kornmayer C, Stahl R, Steinhardt J, Weirich F (2008) Bio-slurries: properties and conditioning. Proceedings of the 16th European biomass conference and exhibition from research to industry and market, Valencia.

70. Solantausta Y, Oasmaa A, Sipilä K (2003) Fast pyrolysis of forestry residues. In: Bridgwater AV (ed) Pyrolysis and gasification of biomass and waste: proceedings of an expert meeting, Strasbourg, September-October 2002, CPL Press, UK. pp 271-276

71. Oasmaa A, Kuoppala E, Gust S, Solantausta Y (2003) Fast pyrolysis of forestry residues. 1. Effect of extractives on phase separation on pyrolysis liquids. Energy & Fuels 17:1-12.

72. Brocksiepe G (1973) Holzverkohlung. In: Bartholome E, Gerhartz W, Ullmann F (eds) Encyklopädie der technischen Chemie, Verlag Chemie, Weinheim, vol 12.703-708

73. Hornung A, Apfelbacher A (2005) Thermo-chemical conversion of straw-haloclean intermediate pyrolysis. Proceedings of the 14th European biomass conference and exhibition biomass for energy, industry and climate protection, Paris.

74. Stahl R, Henrich E, Kögel A, Raffelt K, Steinhardt J, Weirich F, Dinjus E (2004) Pressurized entrained flow gasification of slurries from biomass. In: van Swaaij WPM, Fjällström T, Helm P, Grassi A (eds) Proceedings of the 2nd world conference and technology exhibition on biomass for energy, industry and climate protection, Rome ETA, Florence and WIP, Munich, vol 1.813-816

75. Stahl R, Henrich E, Raffelt K (2009) Pressurized entrained flow gasification of slurries from biomass. Proceedings of ICPS 09, Vienna.

76. Moe JM (1962) Design of water gas shift reactors. Chem Eng Prog 58:33

77. Bonn B (2008) Flugstromvergasung. In: Schmalfeld P (ed) Die Veredelung und Umwandlung von Kohle, DGMK, Hamburg.

78. Bonn B (2008) Shell-Kohlevergasungsverfahren. In: Schmalfeld P (ed) Die Veredelung und Umwandlung von Kohle, DGMK, Hamburg.

79. Haffner S (2010) Modellentwicklung zur numerischen Simulation eines Flugstromvergasers für Biomasse. PhD thesis, Heidelberg University.

80. Rashidi A (2011) CFD Simulation of biomass gasification using detailed chemistry. PhD thesis, Heidelberg University.

81. Seifert H, Kolb T, Leibold H (2009) Syngas aus Biomasse-Flugstromvergasung und Gasreinigung. In: Thome-Kozmiensky KJ (ed) Kraftwerkstechnik: Sichere und nachhaltige Energieversorgung, TK Verlag, Neuruppin.

82. Wang Z, Wang J, Ren F, Han M, Jin Y (2004) Thermodynamics of

the single step synthesis of DME from syngas. Tsinghua Science and Technology 9:169-176

83. Henrich E, Dahmen N, Dinjus E (2009) Cost estimate for biosynfuel production via biosyncrude gasification. Biofuels Bioprod Bioref 3:28-41

84. Trippe F (2009) Techno-ökonomische Prozesskettenanalyse der Schnellpyrolyse als Verfahrensschritt im bioliq® - Konzept. Diploma thesis, Karlsruhe University (TH).

85. Lange S (2008) Systemanalytische Untersuchung zur Schnellpyrolyse als Prozessschritt bei der Produktion von Synthesekraftstoffen aus Stroh und Restholz. PhD thesis. Karlsruhe University (TH).

86. Anderson J (2009) Determining manufacturing costs. Chemical Engineering Progress 105:27-31

87. Onken U, Behr A (1996) Chemische Prozesskunde. Thieme Verlag, Stuttgart.

88. Peters MS, Timmerhaus KD, West RE (2003) Plant design and economics for chemical engineers. McGraw-Hill, New York.

89. Ulrich GD (1984) A guide to chemical engineering process design and economics. Wiley, Hoboken.

90. Peacocke GVC, Bridgwater AV, Brammer JG (2004) Techno-economic assessment of power production from the Wellmann and BTG fast pyrolysis process. In: Bridgwater AV, Boocock DGB (eds) Science in thermal and chemical biomass conversion, CPL Press, UK.

91. Althapp A (2003) Kraftstoffe aus Biomasse mit dem CarboV-Vergasungsverfahren. FVS Fachtagung Regenerative Kraftstoffe, Stuttgart.

92. Rudloff M (2007) First commercial BTL production facility and next generation BTL plants. Proceedings of the synbios conference II, Stockholm.

93. Landàlv I (2005) Status and potential-chemrec black liquor gasification. Proceedings of the synbios conference, Stockholm.

94. Ekbom T, Berglin N, Lindblom M, Ahlvik P (2004) Cost-competitive, efficient bio-methanol/bio-DME production from biomass via black liquor gasification as renewable motor fuels for automotive uses. In: van Swaaij WPM, Fjällström T, Helm

P, Grassi A (eds) Proceedings of the 2nd world conference and technology exhibition on biomass for energy, industry and climate protection, Rome ETA, Florence and WIP, Munich, vol 2.1877-1880

95. Knoef H (ed) (2005) Handbook of biomass gasification. BTG, Enschede

96. Bergmann P, Boersma AR, Kiel JHA, Prins MJ, Ptasinski KJ, Jansen FJ (2004) Torrefaction for entrained flow gasification of biomass. In: van Swaaij WPM, Fjällström T, Helm P, Grassi A (eds) Proceedings of the 2nd world conference and technology exhibition on biomass for energy, industry and climate protection, Rome ETA, Florence and WIP, Munich, vol 1.679-680

Planning for Reliable Coal Quality Delivery Considering Geological Variability: A Case Study in Polish Lignite Mining

Wojciech Naworyta[1], Szymon Sypniowski[2], and Jörg Benndorf[3]

[1]Department of Surface Mining, AGH University of Science and Technology, Mickiewicza Avenue 30, 30-059 Krakow, Poland

[2]Department of Mineral Resources Acquisition, MEERI PAS, Wybickiego Street 7, 31-261 Krakow, Poland

[3]Faculty of Civil Engineering and Geoscience, Delft University of Technology, Building 23, Stevinweg 1, 2600 GA Delft, Netherlands

ABSTRACT

The aim of coal quality control in coal mines is to supply power plants daily with extracted raw material within certain coal quality constraints. On the example of a selected part of a lignite deposit, the problem of quality control for the run-of-mine lignite stream is discussed. The main goal is to understand potential fluctuations and deviations from production targets dependent on design options before an investment is done. A single quality parameter of the deposit is selected for this analysis—the calorific value of raw lignite. The approach requires an integrated analysis of deposit inherent variability, the extraction sequence, and the blending option during material transportation. Based on drill-hole data models capturing of spatial variability of the attribute of consideration are generated. An analysis based on two modelling approaches, Kriging and sequential Gaussian simulation, reveals advantages and disadvantages lead to conclusions about their suitability for the control of raw material quality. In a second step, based on a production schedule, the variability of the calorific value in the lignite stream has been analysed. In a third step the effect of different design options, multiple excavators and a blending bed, was investigated.

INTRODUCTION

Environmental and economic considerations in the electrical energy industry rise the necessity to constantly improve the efficiency of power units. One way to increase the efficiency of energy production in the power plants based on fossil fuels is to supply the raw material with specific and relatively stable quality parameters.

In the case of lignite, the spatial variability of parameters is quite large. Given the variability criterion, lignite belongs to the second group of deposits in the Polish classification. The coefficient of variation v [%] is defined as the ratio of the standard deviation to the mean value of the basic parameters and is usually in the range of 30% to 60%. The exception is the calorific value which has a relatively low volatility in the range of 9 to 16% [1].

To meet customer's requirements, the planning and design of a mining operation have to focus on technical and operational measures to reduce the in situ variability of critical coal attributes during mining and material handling. The aim of different design options, such as the use of blending beds or multiple excavators simultaneously, is to transform the in situ variability in the deposit to a level which meets customers' requirements. For investigating the effect of a coal blending beds the theory of variance reduction in bed blending is well established (e.g., [2]). It is based on the variogram transformation of the incoming to the outgoing stream. Several documented applications (e.g., [3, 4]) use techniques of stochastic simulation based on variograms of critical elements to simulate the variability of incoming material flows and to optimise the transformation process. Considering geologically more complex deposits this approach may be too simplified. To investigate the homogenisation effects in a continuous mining system, the deposit characteristics, in particular the local variability has to be linked with the extraction method, the mining sequence, and blending options throughout the operation [5].

In order to maintain stable raw material parameters, certain measures are undertaken referred to as the lignite stream quality management (e.g., [5–9]). This process begins with the exploration and documentation of the deposit and is conducted until the end of the mine's life. Coal quality control can be divided into several stages:

- identification of critical parameters and modelling of the deposit:
  identification and analysis of critical coal quality parameter,
  spatial modelling of the variability of quality parameters,
- mine planning (long-term planning):
  determination of the location ultimate pit limit and of the opening box cut,
  design of blending options and facilities, such as stock and blending yards,
  establishment of a long-term mining sequence and advances of the mining faces in time,
- exploitation and production control (operational planning):
  short-term production scheduling for the extraction equipment,
  prediction and online analysis of the quality of the extracted coal,
  logistics and transportation,

storage and homogenisation of the raw material.

The analysis presented here relates to the second and third stage of the control process—the design of blending options and operational planning. The following sections will first investigate different geostatistical modelling approaches for their suitability to map realistically spatial variability of lignite attribute considered. In the second part the variability of the extracted material flow is evaluated, including bed-blending and multiple excavators, leading to design options for improved coal quality management and a reliable supply of the power plant.

This paper is a continuation of the aspects related to coal quality management in lignite mines discussed by the authors in previous publications. In particular, methods of conditional simulation in geostatistics investigated in [10, 11] are applied to full scale reserve modelling of a large lignite field in Poland aiming to understand variability of coal quality attributes at a short-term scale. Using these models the second part focuses on design issues of a stock and blending bed to understand its ability to control short-term variation. Contrary to the work described in [5, 12], which focuses on operational optimization of a coal stock and blending bed, here the aim is to understand the effect of the bed size to control coal quality fluctuations of final products to be sold. The combined approach discussed in this paper allows decisions on the optimal stock and blending bed design to be evaluated in the design phase, before short-term operation is actually executed and real fluctuations experienced.

THE OBJECTIVE OF THIS CASE STUDY

For the process of lignite quality control at the stage of operational planning it is necessary to have sufficient exploration information about the deposit. In the mines this task is accomplished in different ways. One of them is to explore the deposit with drill-holes drilled from the roof of the exposed lignite—the so called operational exploration. The holes in the deposit analysed in this paper were drilled in a dense grid of 50 by 50 meters. Although, in comparison to the geological documentation stage, the operational exploration has a higher information content, the actual parameters of the mined lignite still often differ from values identified during this drilling period.

The main objective of this study is to understand possible deviations with respect to the expected calorific of coal produced based on operational exploration data, that is, to assess to what extent these data provide accurate information for the tasks related to the quality control of the mined mineral. To achieve this goal and test the suitability of different approaches, two methods of geostatistical modelling are compared, ordinary Kriging and conditional simulation (e.g., [1]).

In a second step two different design options are investigated focusing on the effect of variability of run-of-mine lignite, which are as follows:

- the availability of a coal stock and blending yard for bed blending: different sizes are investigated,
- the availability of a second excavator and the possibility to blend two lignite streams on the belt conveyor.

For run-of-mine lignite quality control in the context of power plant supply multiple parameters such as calorific value, sulphur content, and silica content have to be taken into account. Without loss of generality, this paper focused on the analysis of the calorific value of the raw lignite Q_i^r.

THE METHOD

On the basis of the operational exploration within the area limits of six-month progress of extraction, variability models of the calorific value in particular mining blocks were created. This analysis was performed for the part of the deposit where the operational exploration is characterized by high regularity. Figure 1 shows the selected part of the deposit with respect to the entire deposit and the assumed mining progress in relation to all exploratory holes.

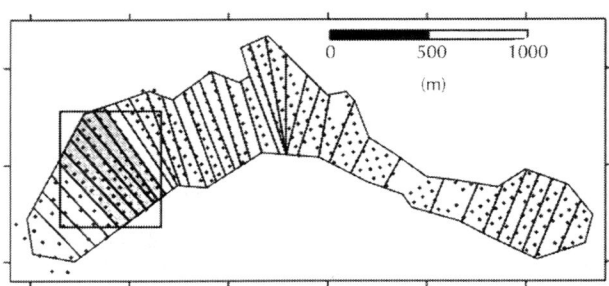

Figure 1: Location of the drill holes of the operational exploration and the limits of monthly mining progress. The rectangle marks the area selected for analysis.

On the basis of the calorific value variability models with a given mining direction, the variations of calorific value were calculated for a six-month period. Figure 2 shows a sequence of mining 195 consecutive exploitation blocks. Each mining block with dimensions of 30 × 30 meters corresponds to an actual average daily production of lignite from the analyzed deposit. With the average lignite seam thickness of approximately 6 meters and a density of 1.15 t/m³, a single exploitation block contains about 6,5 thousand tonnes of lignite.

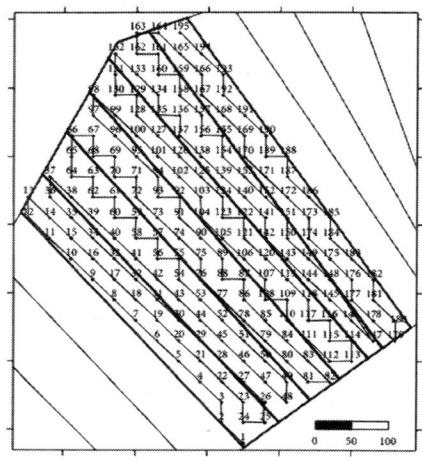

Figure 2: The order of exploitation within a six-month period. Each exploitation block has been marked with the number.

Based on the data the models of spatial variability of the calorific value in the deposit were created using the ordinary Kriging method and the direct sequential simulation method, which are implemented in the software S-GeMS [13]. The geostatistical simulation procedure is based on the idea of Monte-Carlo simulation. Based on available observations of the deposit and on random numbers, the simulation can generate any number of models (herein referred to as realizations). The realizations are unique and at the same time characterized by identical probability to represent the actual deposit. All realizations accurately reflect the values at the observation points. Unlike ordinary Kriging, realizations resulting from simulation accurately reflect the statistical and structural features of the modelled parameters such as the density distribution and spatial variability. Local differences between particular realizations present the measure of uncertainty of the prediction conducted by the simulation on the basis of the available observations. In the paper 50 independent realizations of the calorific value for the selected part of the deposit are presented. Figure 4 shows two exemplifying realizations.

Both of the used methods require a variogram model capturing the spatial variability as input. First an empirical variogram is calculated and secondly a model is fitted. In the presented case the spherical basic structure resulted in the best fit (Figure 3, Table 1). Due to the lack of a clear directional variability in the modelled deposit, an omnidirectional variogram model was used.

Table 1: Basic features of variogram model of calorific value

Variogram model	Dimension and unit
Nugget effect	$160\,000$ $(kJ/kg)^2$
Spherical model	$235\,000$ $(kJ/kg)^2$
Autocorrelation range	$900\,m$

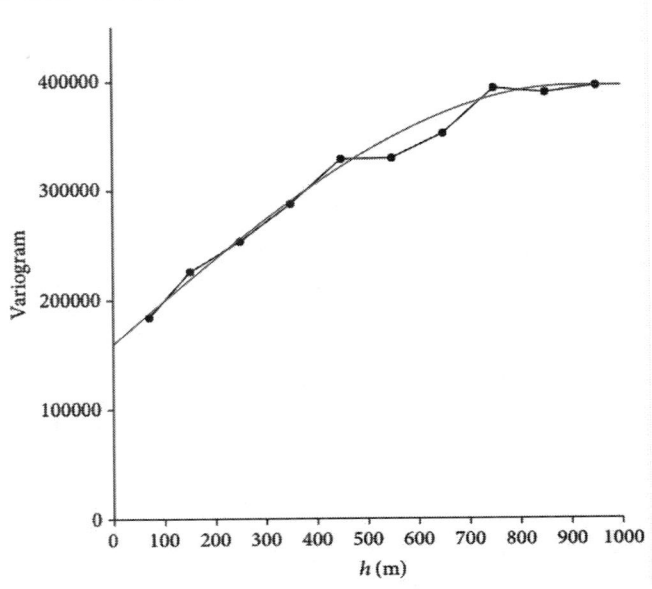

Figure 3: Experimental variogram with variogram model.

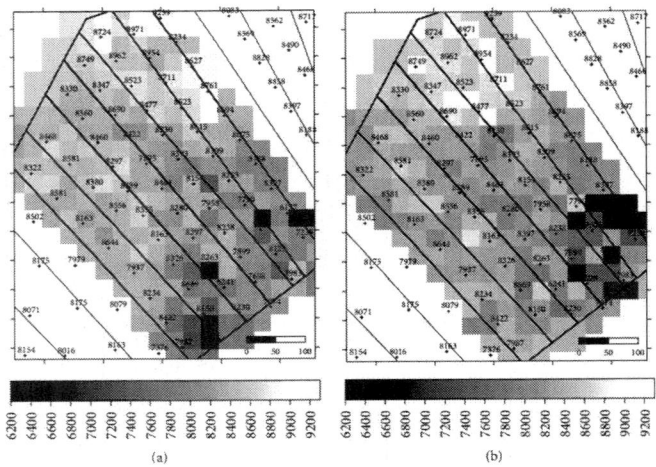

Figure 4: Models of calorific value Q_i^r—exemplifying realizations of geostatistical simulation.

THE DATA BASE USED FOR THE ANALYSIS

The variability models of calorific value were created based on 68 operational exploration drill holes located within the borders of mining and on the basis of the adjacent holes. Table 2 summarizes the basic statistical characteristics of the measurement data from 68 holes. As can be seen, both models perform well in reproducing the mean value of the drill holes. The variance of modelled blocks cannot directly be compared to the variance of exploration data, since both are based on a different support. However, it can be noticed that simulated block values appear more variable as Kriged block values. This effect results from the smoothing effect of Kriging.

Table 2: Basic statistics of calorific value based on 68 boreholes of the operational exploration and of the both models

	Data from the exploratory holes	Model ordinary Kriging	Model-exemplifying simulation
Number of holes/number of blocks	68	195 (30×30 m)	195 (30×30 m)
The mean value	8267 kJ/kg	8274 kJ/kg	8260 kJ/kg
The standard deviation	521 kJ/kg	316 kJ/kg	485 kJ/kg
Coefficient of variation	6,30%	3,80%	5,9%
The minimum value	6137 kJ/kg	7517 kJ/kg	6225 kJ/kg
The maximum value	9259 kJ/kg	8759 kJ/kg	9006 kJ/kg

RESULTS AND DISCUSSION

Figures 4 and 5 show the calorific value volatility models in the selected part of the deposit. To facilitate the assessment of the validity of the models, the figures also present the location of the operational exploration holes with their identified calorific value. Figure 4 shows two examples out of the total 50 conducted realizations of the simulation. The models differ from each other, and the differences are primarily in the blocks where there are no exploratory drill holes.

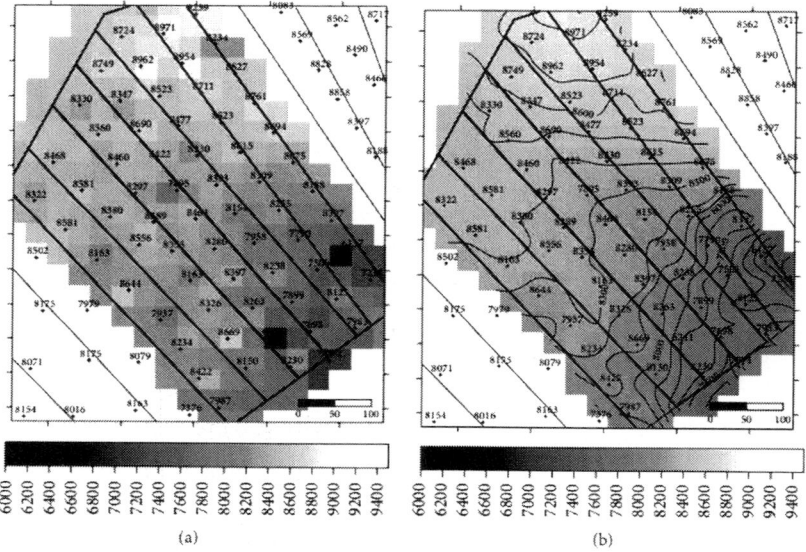

(a)   (b)

Figure 5: Variation of the calorific value Q_i^r average of 50 realizations of the simulation (l) and ordinary Kriging (r).

Figure 5 summarizes the two models representing the expected spatial distribution of the calorific value—the average of the 50 realizations (a) and using the ordinary Kriging method (b). In the model created using the Kriging the calorific value changes gradually. The contour lines shown in Figure 5(b) show the effect of smoothing that occurred when using ordinary Kriging. Contrarily the variability is conserved in the single realization (Figure 4). When averaging all realizations, resulting in the so-called E-type estimator (Figure5(a)), a very similar model to the one of Kriging is obtained.

Figure 6 presents the histograms of the calorific value models variability in the selected part of the deposit. There is an apparent narrowing of values in the ordinary Kriging model. Note that the Kriging smoothing effect can be compensated by implementing the Yamamoto correction.

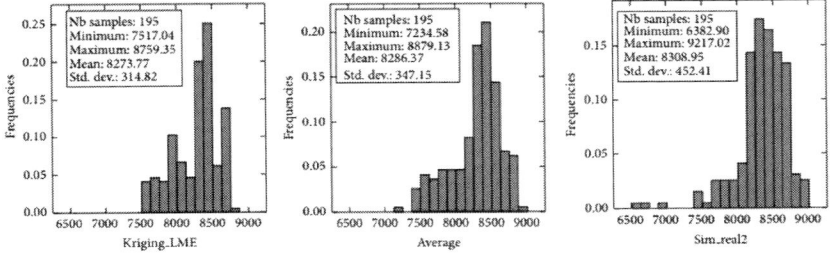

Figure 6: Histograms of calorific value Q_i^r based on ordinary Kriging model (l), average of 50 realizations of a geostatistical simulation (m), and one realization of a geostatistical simulation (r).

Figure 7(a) shows the standard deviation of ordinary Kriging, which expresses the magnitude of the expected interpolation error. Its size in any given block depends primarily on the distance to the nearest observation, on the basis of which the interpolation was conducted. This relation results mainly from the variogram and the data configuration.

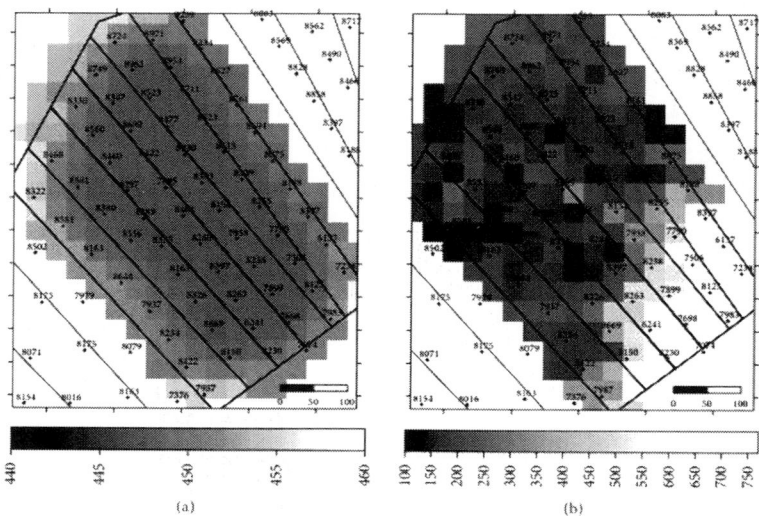

(a)   (b)

Figure 7: Standard deviation of Kriging (l) and standard deviation of simulation based on 50 realizations (r).

Kriging's standard deviation is independent of local variation of observations used for modelling. Figure7(b) presents the map of the conditional simulation's standard deviation. The map is a result of a statistical analysis of 50 realizations. In each node of the grid standard deviation was calculated, reflecting the uncertainty of a local forecast. There are some clear differences between the two figures. These differences appear not only in the nominal value of the standard deviation, but also in their spatial distribution in the modelled deposit. The standard deviation of the simulation shows particularly high values in the south-eastern part of the deposit. This is the influence of high calorific value variation of the adjacent observations.

Based on the assumed extraction schedule (Figure 2), graphs of the calorific value in the subsequently mined blocks (corresponding to the average daily production volumes) were prepared. Figures 8 and 9 depict the variations of the calorific value in the lignite stream during six months of mining. The graph in Figure 8 was created using the variability model prepared with the use of the ordinary Kriging method. Besides the mean value, the dotted lines constitute for the Kriging's standard deviation of the respective exploitation blocks. The graph in Figure 9 shows the variation of the calorific value based on the model created by conditional simulation method. Three exemplifying realizations of the simulation are shown together with the mean of all 50 realizations. Graphs (Figures 8 and 9) are supplemented with horizontal lines corresponding to the average value calculated from 68 observations (Q_{sr} = 8267 kJ/kg) and the lines corresponding to the average increased (Q_g = 8788 kJ/kg) and the average decreased by the value of the standard deviation of the observation (Q_d = 7746 kJ/kg).

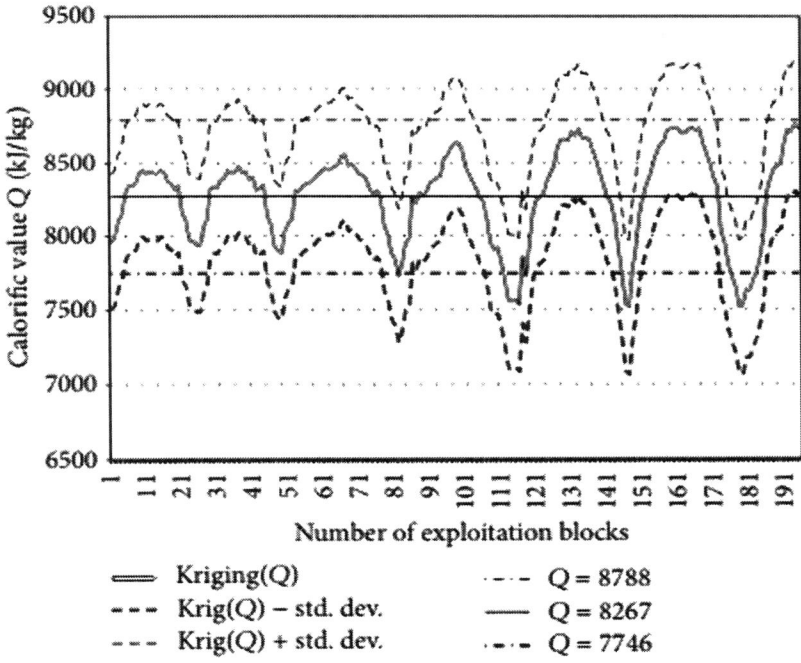

Figure 8: The fluctuation of calorific value Q_i^r within 195 days of exploitation—based on ordinary Kriging.

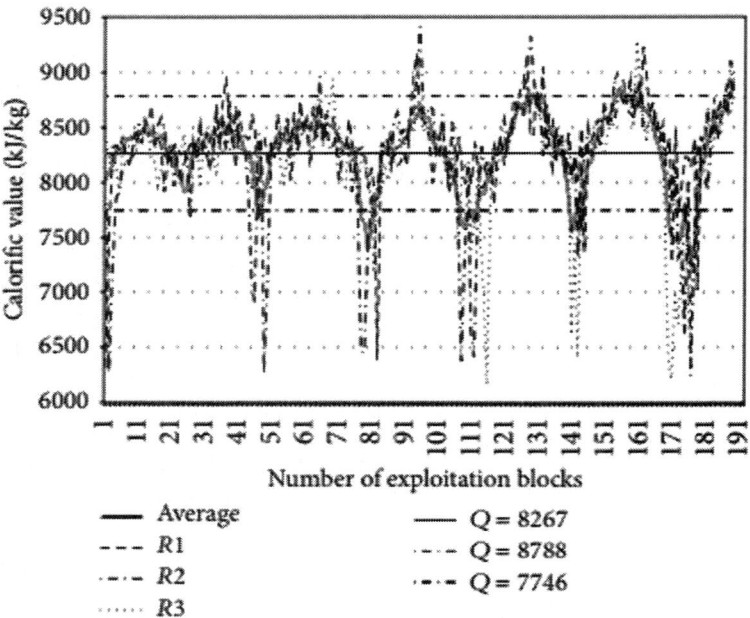

Figure 9: The fluctuation of calorific value Q_i^r within 195 days of exploitation—based on conditional simulation.

In the first graph (Figure 8), the average calorific value determined using Kriging changes cyclically within the range of the standard deviation of observation, extending only slightly beyond those lines. In the second graph (Figure 9) the volatility of individual realization (R1, R2, and R3) is significant, and the mean of realizations (average R1, R1, ..., R50) in several places goes far beyond the limits of the line marking Q_d=7746kJ/kg, reaching a value below Q=6500kJ/kg.

Based on the results of the 50 realizations of simulation, a map showing the probability of exceeding the thresholds established in the particular exploitation blocks was created. The values of the mean of 68 observations plus and minus the standard deviation were chosen as the assumed limits (thresholds), which is rounded, respectively, to Q_g=8790kJ/kg and Q_d=7750kJ/kg. Figure 10 shows the map of the probability of exceeding the adopted thresholds.

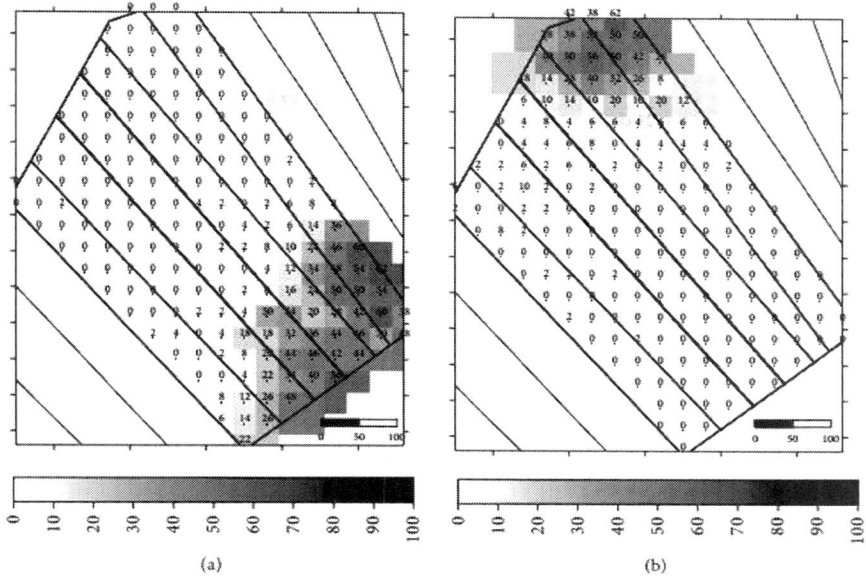

Figure 10: The occurrence probability of calorific value lower than Q_d=7750kJ/kg (l) and higher than Q_g=8790 kJ/kg (r) in the exploitation blocks, based on simulation.

INVESTIGATION OF DESIGN OPTIONS

Option 1 (bed blending using a coal-stock-and-blending yard). Bed blending has three objectives: namely, buffering, composing, and homogenising. Thereby it transforms the characteristics of the incoming material flow in an outgoing material flow, whose characteristics are defined by costumer specifications and may be of contractual relevance. The characteristics of the incoming material flow are a function of the geological conditions, the applied selectivity in extracting the deposit, the mining sequence, and the operation mode in the pit as discussed in the previous section. The following considerations concern the homogenisation effect of using bed blending. The efficiency of blending and smoothing variability is significantly dependent on constructive parameters as well as the operation of the blending yard. Constructive factors are the type of the yard, its length and width, the angle of repose, the number of layers, and speed of the stacker. The following constructive parameters are given in the case study: the blending yard is of type "strata" (Figure11).

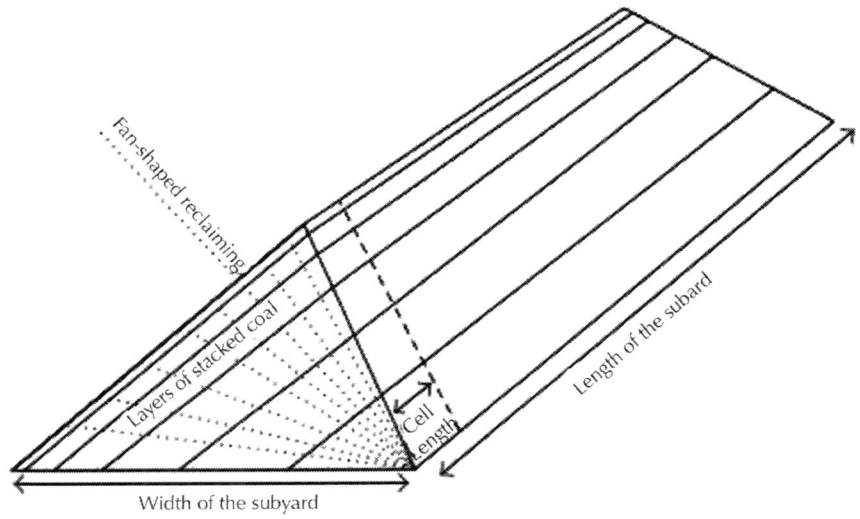

Figure 11: Schematic illustration of a strata-type blending yard.

The coal coming from the pit (incoming material flow) is stacked into layers, which are spread along the total length of a bed by a continuously up and down moving stacker. The number and thickness of the several layers are variable and can be influenced by the moving rate of the stacker dependent on the total production rate of the mine. At maximum about 61 layers can be placed in a pocket. The yard is reclaimed in a fan-shaped manner orthogonally to the alignment of the stacked layers by a scraper. In this way the coal quality of the outgoing material flow is formed as an average over the total number of stacked layers.

Investigations have shown that operating with >15 layers the incoming flow can be completely homogenized [5, 12]. Therefore in this investigation it is assumed that the homogenisation effect is solely dependent on the stockpile size. Figure 12 shows the variability of the outgoing material flow for the different blending yard sizes: 0 kt, 60 kt, 180 kt, and 300 kt. Clearly already a considerable small blending bed size leads to a significant homogenization. Considering the already previously introduced lower and upper limit of Q_g=8790 kJ/kg and Q_d=7750 kJ/kg it would need a stockpile size of >300 kt to ensure continuously in-spec delivery of the power plant.

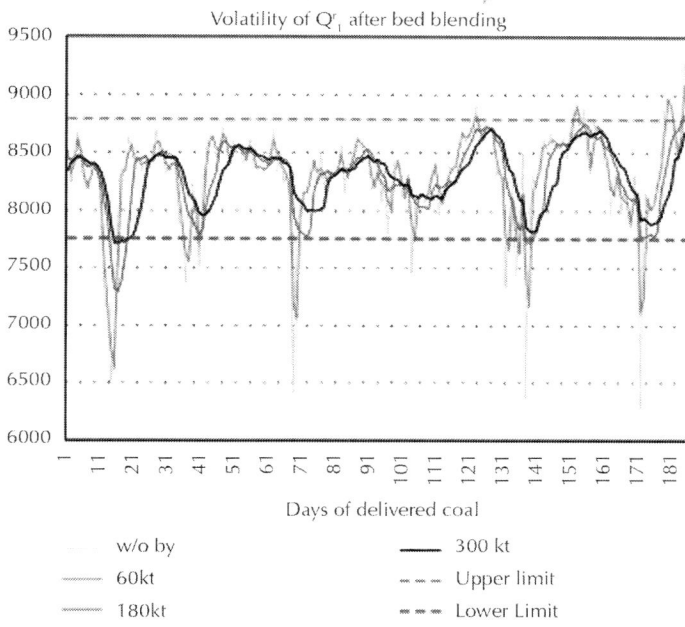

Figure 12: The fluctuation of calorific value Q_i^r within 195 days of exploitation—after bed blending.

Figure 13 shows a summary of the frequencies of expected deviations from production targets for different blending bed sizes. For example a size of 180 kt would still lead to approximately 5% of daily deliveries deviating from potentially contractually fixed limits. A size of 330 kt would ensure that the in situ variability of the deposit can be transformed into a product exhibiting a maximum variability as requested from the customer. In addition this size of a stock pile would form a buffer bridging about 11 days of production and can ensure continuous supply of the power plant during small and medium termed maintenance or breakdown events.

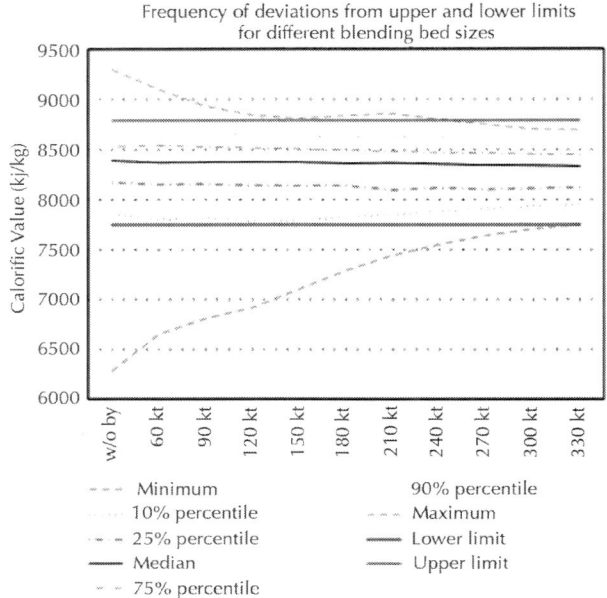

Figure 13: Distribution of calorific value as a function of blending yard sizes.

Option 2 (availability of two excavators). This design option considers the availability of two excavators, which are operated simultaneously. For example excavator one may excavate the first part of the bench to the middle and excavator two extracts the remaining blocks. To avoid installed overcapacity the capacity of each of the two excavators can be designed as low as half of the capacity of on single excavator achieving the same daily production target of 30 kt. For this investigation it was assumed that both excavators operate at an extraction rate of 15 kt per day. Assuming no targeted quality optimized scheduling, which means both excavators are always operating at one half of the bench without pinpointed schedule, Figure 14 shows the result of the blended stream of lignite. As can be seen, a simultaneous extraction of blocks with a subsequent blending on the belt conveyor significantly reduces the variability. Considering the coal quality production limits it becomes obvious that there may still occur sporadic deviations from production targets. These can be avoided by quality optimized scheduling or using an additional blending yard with a small capacity, for example, 60 kt.

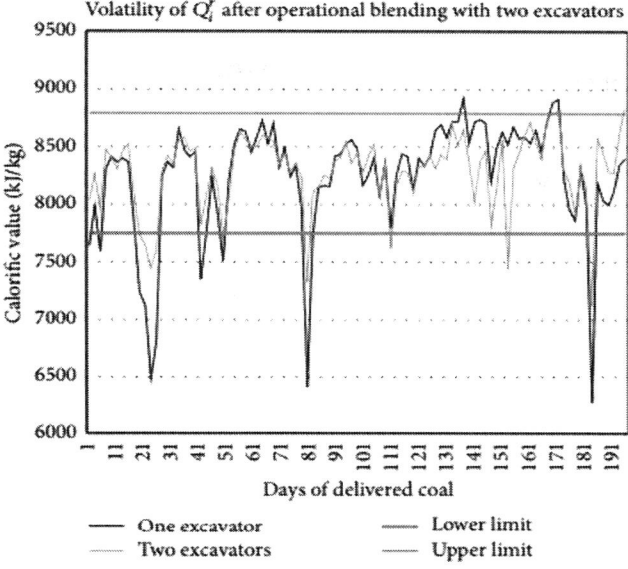

Figure 14: The fluctuation of calorific value Q_i^r within 195 days of exploitation—using two excavators simultaneously.

CONCLUSIONS

The calorific value of the analysed part of the deposit has a relatively low volatility (6,3%), yet due to the elongated shape of the deposit (Figure 1), which implies the direction of mining and distribution of the calorific value, the average daily calorific values are in the range of 7750–8790 kJ/kg.

With the accepted method of mining, changes occur in almost regular monthly cycles. In two parts of the deposit the lignite has wider-than-threshold values (Figure 10). This applies mainly to the values lower than 7750 kJ/kg in the south-eastern area, as well as more than 8790 kJ/kg in the northern part of the deposit. For purposes of coal quality control in order to maintain the calorific value at the desired level, it is useful to apply modern spatial interpolation tools. The study shows that for this purpose geostatistical simulation is particularly useful as it—in addition to the mean values—allows determining the level of the probability of exceeding the adopted thresholds in the particular blocks (risk level). In contrast to the simulation, using the ordinary Kriging interpolation may lead to erroneous operating decisions because of the effect of smoothing of the extreme values demonstrated in the paper.

The property of geostatistical simulation to reproduce in situ variability can be used to investigate the variability in dependence of certain design options in the subsequent material handling system. In the present case the availability and size of blending beds were investigated as well as the availability of an additional excavator. It has been shown that both options can contribute significantly to the reduction of variability in CV. In addition a required stock-pile size could be defined that ensures a continuous in-spec delivery of coal to the customer.

REFERENCES

1. W. Naworyta and S. Sypniowski, "About the problem of lignite stream quality control in the context of proper identification of deposit's quality parameters," Surface Mining, no. 2, pp. 58–65, 2013 (Polish).

2. P. M. Gy, "A new theory of bed-blending derived from the theory of sampling—development and full-scale experimental check," International Journal of Mineral Processing, vol. 8, no. 3, pp. 201–238, 1981

3. M. Kumral, "Bed blending design incorporating multiple regression modelling and genetic algorithms," The Journal of the Southern African Institute of Mining and Metallurgy, vol. 106, no. 3, pp. 229–236, 2006

4. D. Marques, J. F. Costa, D. Ribeiro, and J. C. Koppe, "The evidence of volume variance relationship in blending and homogenisation piles using stochastic simulation," in Proceedings of the 4th World Forum on Sampling and Blending, pp. 235–242, The Southern African Institute of Mining and Metallurgy, 2009.

5. J. Benndorf, "Investigating the variability of key coal quality parameters in continuous mining operations when using stockpiles," in Advances in Orebody Modelling and Strategic

Mine Planning I, AusIMM, 2011.

5.   J. Benndorf, "Investigating in situ variability and homogenisation of key quality parameters in continuous mining operations," Transactions of the Institutions of Mining and Metallurgy, Section A: Mining Technology, vol. 122, no. 2, pp. 78–85, 2013

7.   D. Gärtner and R. Hempel, Monitoring and Control of Processes in the Lignite Mines in Rhineland, Lignite Mining, Springer, Heidelberg, Germany, 2009.

8.   L. Kunde and D. Trummer, "Coal quality management, lignite mining—Kohlenqualitätsmanagement (germ.)," in Der Braunkohlentagebau, pp. 409–426, Springer, Berlin, Germany, 2009.

9.   B. Zimmer, "Development of a new on-line coal quality management system in a lignite mine in Serbia," in Continuous Surface Mining, Latest Development in Mine Planning, Equipment and Environmental Protection: Proceedings of 10th International Symposium Continuous Surface Mining, 13–15 September 2010, C. Drebenstedt, Ed., pp. 290–302, Technische Universitat Bergakademie Freiberg, Freiberg, Germany, 2010.

10.   W. Naworyta, "Variability analysis of lignite deposit parameters for output quality control," Mineral Resources Management, vol. 24, no. 2–4, pp. 97–110, 2008 (Polish).

11.   W. Naworyta and J. Benndorf, "Accuracy assessment of geostatisticalmodelling methods of mineral deposits for the purpose of their future exploitation—based on one lignite deposit," Mineral Resources Management, vol. 28, no. 1, pp. 77–101, 2012 (Polish).

12.   J. Benndorf, "Application of efficient methods of conditional simulation for optimising coal blending strategies in large continuous open pit mining operations," International Journal of Coal Geology, vol. 112, pp. 141–153, 2013.

13.   N. Remy, A. Boucher, and J. Wu, Applied Geostatistics with S-GeMS, Cambridge University Press, Cambridge, UK, 2009.

Citations

CHAPTER 1

Andrea Giubergia, Daniel Riesco, Verónica Gil-Costa, and Marcela Printista, "UML Profile for Mining Process: Supporting Modeling and Simulation Based on Metamodels of Activity Diagram," Modelling and Simulation in Engineering, vol. 2014, Article ID 974850, 10 pages, 2014. doi:10.1155/2014/974850.

CHAPTER 2

Björn **Öhlander** , Terrence Chatwin and Lena Alakangas, Management of Sulfide-Bearing Waste, a Challenge for the Mining Industry, doi:10.3390/min2010001.

CHAPTER 3

Gergely Tóth, Ciro Gardi, Katalin Bódis, **Éva** Ivits, Ece Aksoy, Arwyn Jones, Simon Jeffrey, Thorum Petursdottir, and Luca Montanarella, Continental-Scale Assessment of Provisioning Soil Functions in Europe, doi:10.1186/2192-1709-2-32.

CHAPTER 4

P. Koltun and A. Tharumarajah, "Life Cycle Impact of Rare Earth Elements," ISRN Metallurgy, vol. 2014, Article ID 907536, 10 pages, 2014. doi:10.1155/2014/907536.

CHAPTER 5

Elnaz Ghadimi, Hazem Eimar, Benedetto Marelli, Showan N Nazhat Masoud Asgharian, Hojatollah Vali, and Faleh Tamimi, Trace elements can influence the physical properties of tooth enamel, doi:10.1186/2193-1801-2-499.

CHAPTER 6

Larry M Deschaine, Theodore P Lillys, and János D Pintér, Groundwater Remediation Design Using Physics-Based Flow, Transport, and Optimization Technologies, doi:10.1186/2193-2697-2-6.

CHAPTER 7

Nicolaus Dahmen, Edmund Henrich, Eckhard Dinjus, and Friedhelm Weirich, The Bioliq Bioslurry Gasification Process for the Production of Biosynfuels, Organic Chemicals, and Energy, doi:10.1186/2192-0567-2-3.

CHAPTER 8

Wojciech Naworyta, Szymon Sypniowski, and Jörg Benndorf, "Planning for Reliable Coal Quality Delivery Considering Geological Variability: A Case Study in Polish Lignite Mining," Journal of Quality and Reliability Engineering, vol. 2015, Article ID 941879, 9 pages, 2015. doi:10.1155/2015/941879.

Index